SEED TECHNOLOGY AND SEED PATHOLOGY

SEED TECHNOLOGY AND SEED PATHOLOGY

Uma Shankar Singh

ANMOL PUBLICATIONS PVT. LTD.
NEW DELHI - 110 002 (INDIA)

ANMOL PUBLICATIONS PVT. LTD.
H.O.: 4374/4B, Ansari Road, Daryaganj,
New Delhi-110 002 (India)
Ph.: 23278000, 23261597
B.O.: No. 1015, Ist Main Road, BSK IIIrd Stage
IIIrd Phase, IIIrd Block,
Bangalore - 560 085 (India)
Visit us at: www.anmolpublications.com

Seed Technology and Seed Pathology

ISBN 978-81-261-3675-9

PRINTED IN INDIA

Printed at Mehra Offset Press, Delhi.

Contents

Preface

The production and supply of good quality seed of high-yielding varieties is one of the most important requirements of modern agriculture and a challenge to seed technologists. Seeds play an important role in disseminating pathogenic organisms to areas from which, they have been absent and seed-borne infection may be the starting point for a serious disease in the growing crop. During seed storage, the metabolic activities of microorganisms may cause even greater losses. Hence, an integrated disease management approach and suitable seed technology needs to be developed.

This book, *Seed Technology and Seed Pathology,* is a unique, comprehensive book and provides up-to-date information relating to recent areas of research in Seed Technology and Seed Pathology in the Indian perspective. This book is extremely useful for teachers, researchers and students in the field of Seed Science and Technology, Seed Pathology, Plant Pathology, Microbiology, Horticulture, Forestry, Entomology and others interested in different branches of Botany and Agriculture."

Author

Chapter 1

Introduction

A seed is a small embryonic plant enclosed in a covering called the seed coat, usually with some stored food. It is the product of the ripened ovule of gymnosperm and angiosperm plants which occurs after fertilization and some growth within the mother plant. The formation of the seed completes the process of reproduction in seed plants (started with the development of flowers and pollination), with the embryo developed from the zygote and the seed coat from the integuments of the ovule.

Seeds have been an important development in the reproduction and spread of flowering plants, relative to more primitive plants like mosses, ferns and liverworts, which do not have seeds and use other means to propagate themselves. This can be seen by the success of seed plants (both gymnosperms and angiosperms) in dominating biological niches on land, from forests to grasslands both in hot and cold climates.

The term *seed* also has a general meaning that predates the above — anything that can be sown i.e. "seed" potatoes, "seeds" of corn or sunflower "seeds". In the case of sunflower and corn "seeds", what is sown is the seed enclosed in a shell or hull, and the potato is a tuber.

SEED STRUCTURE

A typical seed includes three basic parts:

1. An embryo,

2. A supply of nutrients for the embryo,

3. A seed coat.

The embryo is an immature plant from which a new plant will grow under proper conditions. The embryo has one cotyledon or seed leaf in monocotyledons, two cotyledons in almost all dicotyledons and two or more in gymnosperms. The radicle is the embryonic root. The plumule is the embryonic shoot. The embryonic stem above the point of attachment of the cotyledon(s) is the epicotyl. The embryonic stem below the point of attachment is the hypocotyl.

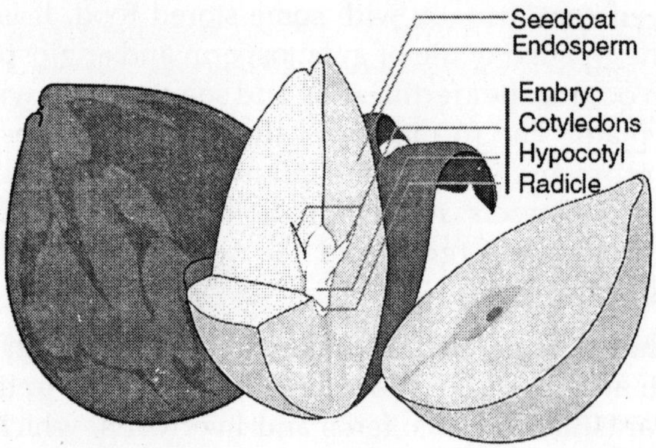

Fig. The parts of an avocado seed (a dicot), showing the seed coat, endosperm, and embryo.

Within the seed, there usually is a store of nutrients for the seedling that will grow from the embryo. The form of the stored nutrition varies depending on the kind of plant. In angiosperms, the stored food begins as a tissue called the endosperm, which is derived from the parent plant via double fertilization. The usually triploid endosperm is rich in oil or starch and protein.

In gymnosperms, such as conifers, the food storage tissue is part of the female gametophyte, a haploid tissue. In some species, the embryo is embedded in the endosperm or female gametophyte, which the seedling will use upon germination.

In others, the endosperm is absorbed by the embryo as the latter grows within the developing seed, and the cotyledons of the embryo become filled with this stored food.

At maturity, seeds of these species have no endosperm and are termed exalbuminous seeds. Some exalbuminous seeds are bean, pea, oak, walnut, squash, sunflower, and radish. Seeds with an endosperm at maturity are termed albuminous seeds.

Most monocots (e.g. grasses and palms) and many dicots (e.g. brazil nut and castor bean) have albuminous seeds. All gymnosperm seeds are albuminous.

The seed coat (or *testa*) develops from the tissue, the integument, originally surrounding the ovule. The seed coat in the mature seed can be a paper-thin layer (e.g. peanut) or something more substantial (e.g. thick and hard in honey locust and coconut). The seed coat helps protect the embryo from mechanical injury and from drying out.

In addition to the three basic seed parts, some seeds have an appendage on the seed coat such an aril (as in yew and nutmeg) or an elaiosome (as in Corydalis) or hairs (as in cotton). There may also be a scar on the seed coat, called the *hilum*; it is where the seed was attached to the ovary wall by the funiculus.

SEED PRODUCTION

Seeds are produced in several related groups of plants, and their manner of production distinguishes the angiosperms ("enclosed seeds") from the gymnosperms ("naked seeds"). Angiosperm seeds are produced in a hard or fleshy (or with layers of both) structure called a fruit that encloses the seeds, hence the name.

In gymnosperms, no special structure develops to enclose the seeds, which begin their development "naked" on the bracts of cones. However, the seeds do become covered by the cone scales as they develop in some species of conifer.

Kinds of Seeds

There are a number of modifications to seeds by different groups of plants. One example is that of the so-called *stone* fruits (such as the peach), where a hardened fruit layer (the endocarp) surrounds the actual seed and is fused to it.

Many structures commonly referred to as "seeds" are actually dry fruits. Sunflower seeds are sold commercially while still enclosed within the hard wall of the fruit, which must be split open to reach the seed.

Seed development

The inside of a *Ginkgo* seed, showing a well-developed embryo, nutritive tissue (megagametophyte), and a bit of the surrounding seed coat.

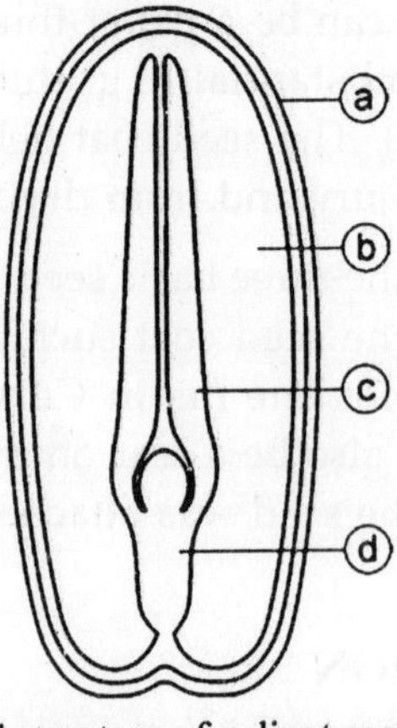

Fig. Diagram of the internal structure of a dicot seed and embryo. (a) seed coat, (b) endosperm, (c) cotyledon, (d) hypocotyl.

The seed, which is an embryo with two points of growth (one of which forms the stems the other the roots) is enclosed in a seed coat with some food reserves. Angiosperm seeds consist of three genetically distinct constituents:

1. The embryo formed from the zygote,
2. The endosperm, which is normally triploid,
3. The seed coat from tissue derived from the maternal tissue of the ovule.

In angiosperms, the process of seed development begins with double fertilization and involves the fusion of the egg and sperm nuclei into a zygote.

The second part of this process is the fusion of the polar nuclei with a second sperm cell nucleus, thus forming a primary endosperm. Right after fertilization the zygote is mostly inactive but the primary endosperm divides rapidly to form the endosperm tissue. This tissue becomes the food that the young plant will consume until the roots have developed after germination or it develops into a hard seed coat.

The seed coat forms from the two integuments or outer layers of cells of the ovule, which derive from tissue from the mother plant, the inner integument forms the tegmen and the outer forms the testa. When the seed coat forms from only one layer it is also called the testa, though not all such testa are homologous from one species to the next.

In gymnosperms, the two sperm cells transferred from the pollen do not develop seed by double fertilization but instead only one sperm fertilizes the egg while the other is not used. The seed is composed of the embryo (the result of fertilization) and tissue from the mother plant, which also form a cone around the seed in coniferous plants like Pine and Spruce.

The ovules after fertilization develop into the seeds; the main parts of the ovule are the funicle; which attaches the ovule to the placenta, the nucellus; the main region of the ovule were the embryo sac develops, the micropyle; A small pore or opening in the ovule where the pollen tube usually enters during the process of fertilization, and the chalaza; the base of the ovule opposite the micropyle, where integument and nucellus are joined together.

The shape of the ovules as they develop often affects the finale shape of the seeds. Plants generally produce ovules of four shapes: the most common shape is called anatropous, with a curved shape. Orthotropous ovules are straight with all the parts of the ovule lined up in a long row producing an uncurved seed. Campylotropous ovules have a curved embryo sac often giving the seed a tight "c" shape. The last ovule shape is called amphitropous, where the ovule is partly inverted and turned back 90 degrees on its stalk or funicle.

In the majority of flowering plants the zygotes first division is transversely orientated in regards to the long axis and this establishes the polarity of the embryo. The upper or chalazal pole becomes the main area of growth of the embryo, while the lower or micropylar pole produces the stalk-like suspensor that attaches to the micropyle. The suspensor absorbs and manufacturers nutrients from the endosperm that are utilized during the embryos growth.

The embryo is composed of different parts; the epicotyle will grow into the shoot, the radicle grows into the primary root, the hypocotyl connects the epicotyle and the radicle, the cotyledons form the seed leaves, the testa or seed coat forms the outer covering of the seed. Monocotyledonous plants like corn, have other structures; instead of the hypocotyle-epicotyle, it has a coleoptile that forms the first leaf and connects to the coleorhiza that connects to the primary root and adventitious roots form from the sides.

The seeds of corn are constructed with these structures; pericarp, scutellum (single large cotyledon) that absorbs nutrients from the endosperm, endosperm, plumule, radicle, coleoptile and coleorhiza - these last two structures are sheath-like and enclose the plumule and radicle, acting as a protective covering. The testa or seed coats of both monocots and dicots are often marked with patterns and textured markings, or have wings or tufts of hair.

Seed size and Seed Set

Seeds are very diverse in size. The dust-like orchid seeds are the smallest with about one million seeds per gram. Embryotic seeds have immature embryos and no significant energy reserves. They are myco-heterotrophs, depending on mycorrhizal fungi for nutrition during germination and the early growth of the seedling, in fact some terrestrial Orchid seedlings spend the first few years of their life deriving energy from the fungus and do not produce green leaves. At over 20 kg, the largest seed is the coco de mer.

Plants that produce smaller seeds can generate many more seeds while plants with larger seeds invest more resources into those seeds and normally produce fewer seeds. Small seeds are quicker to ripen and can be dispersed sooner, so fall blooming plants often have small seeds. Many annual plants produce great quantities of smaller seeds; this helps to ensure that at least a few will end in a favorable place for growth. Herbaceous perennials and woody plants often have larger seeds, they can produce seeds over many years, and larger seeds have more energy reserves for germination and seedling growth and produce larger, more established seedlings.

SEED FUNCTIONS

Seeds serve several functions for the plants that produce them. Key among these functions are nourishment of the embryo, dispersal to a new location, and dormancy during unfavorable conditions. Seeds fundamentally are a means of reproduction and most seeds are the product of sexual reproduction which produces a remixing of genetic material and phenotype variability that natural selection acts on.

Embryo Nourishment

Seeds protect and nourish the embryo or baby plant. Seeds usually give a seedling a faster start than a sporling from a spore gets because of the larger food reserves in the seed.

Seed Dispersal

Unlike animals, plants are limited in their ability to seek out favorable conditions for life and growth. As a consequence, plants have evolved many ways to disperse their offspring by dispersing their seeds. A seed must somehow "arrive" at a location and be there at a time favorable for germination and growth.

When the fruits open and release their seeds in a regular way, it is called dehiscent, which is often distinctive for related groups of plants, these fruits include; Capsules, follicles, legumes, silicles and siliques. When fruits do not open and release their seeds in a regular fashion they are called indehiscent, which include these fruits; Achenes, caryopsis, nuts, samaras, and utricles.

Seed dispersal is seen most obviously in fruits; however many seeds aid in their own dispersal. Some kinds of seeds are dispersed while still inside a fruit or cone, which later opens or disintegrates to release the seeds. Other seeds are expelled or released from the fruit prior to dispersal. Milkweeds produce a fruit type, known as a *follicle,* that splits open along one side to release the seeds. Iris capsules split into three "valves" to release their seeds.

By Wind (Anemochory)

Dandelion seeds (achenes) can be carried long distances by the wind.

- Many seeds (e.g. maple, pine) have a wing that aids in wind dispersal.
- The dustlike seeds of orchids are carried efficiently by the wind.
- Some seeds, (e.g. dandelion, milkweed, poplar) have hairs that aid in wind dispersal.

By Water (hydrochory)

- Some plants, such as *Mucuna* and *Dioclea,* produce buoyant seeds termed sea-beans or drift seeds because they float in rivers to the oceans and wash up on beaches.

By Animals (Zoochory)

- Seeds (burrs) with barbs or hooks (e.g. acaena, burdock, dock which attach to animal fur or feathers, and then drop off later.
- Seeds with a fleshy covering (e.g. apple, cherry, juniper) are eaten by animals (birds, mammals) which then disperse these seeds in their droppings.

- Seeds (nuts) which are an attractive long-term storable food resource for animals (e.g. acorns, hazelnut, walnut); the seeds are stored some distance from the parent plant, and some escape being eaten if the animal forgets them.

Myrmecochory

Myrmecochory is the dispersal of seeds by ants. Foraging ants disperse seeds which have appendages called elaiosomes (e.g. bloodroot, trilliums, Acacias, and many species of Proteaceae). Elaiosomes are soft, fleshy structures that contain nutrients for animals that eat them. The ants carry such seeds back to their nest, where the elaiosomes are eaten.

The remainder of the seed, which is hard and inedible to the ants, then germinates either within the nest or at a removal site where the seed has been discarded by the ants. This dispersal relationship is an example of mutualism, since the plants depend upon the ants to disperse seeds, while the ants depend upon the plants seeds for food. As a result, a drop in numbers of one partner can reduce success of the other.

In South Africa, the Argentine ant (*Linepithema humile*) has invaded and displaced native species of ants. Unlike the native ant species, Argentine ants do not collect the seeds of *Mimetes cucullatus* or eat the elaiosomes. In areas where these ants have invaded, the numbers of *Mimetes* seedlings have dropped.

Seed Dormancy and Protection

One important function of most seeds is delaying germination, which allows time for dispersal and prevents germination of all the seeds at one time when conditions appear favorable. The staggering of germination safeguards some seeds or seedlings from suffering during short periods of bad weather, transient herbivores or competition from other plants for light and nutrients. Many species of plants have seeds that germinate over many months or years, and some seeds can remain in the soil seed bank for more than 50 years before germination.

Seed dormancy is defined as a seed failing to germinate under environmental conditions optimal for germination, normally when the seed's environment is at the right temperature with proper soil moisture conditions. Induced dormancy or seed quiescence occurs when a seed fails to germinate because the external environmental conditions are inappropriate for germination, mostly in response to being too cold or hot, or too dry.

True dormancy or innate dormancy is caused by conditions within the seed that prevent germination under normally ideal conditions. Often seed dormancy is divided into four major categories: exogenous; endogenous; combinational; and secondary.

Exogenous dormancy is caused by conditions outside the embryo including:

- Hard seed coats or physical dormancy occurs when seeds are impermeable to water or the exchange of gases. In some seeds the seed coat physically prevents the seedling from growing.
- Chemical dormancy includes growth regulators etc.

Endogenous dormancy is caused by conditions within the embryo itself, including:

- Immature embryos where some plants release their seeds before the tissues of the embryos have fully differentiated, and the seeds ripen after they take in water while on the ground, germination can be delayed from a few weeks to a few months.
- Morphological dormancy where seeds have fully differentiated embryos that need to grow more before seed germination, the embryos are not yet fully developed.
- Morphophysiological dormancy seeds with underdeveloped embryos, and in addition have physiological components to dormancy. These seeds therefore require a dormancy-breaking treatments as

well as a period of time to develop fully grown embryos.

- Physiological dormancy prevents seed germination until the chemical inhibitors are broken down or are no longer produced by the seed, often physiological dormancy is broken by a period of cool moist conditions, normally below (+4C) 39F, or in the case of many species in *Ranunculaceae* and a few others,(-5C) 24F. Other chemicals that prevent germination are washed out of the seeds by rainwater or snow melt. Abscisic acid is usually the growth inhibitor in seeds and its production can be affected by light. Some plants like Peony species have multiple types of physiological dormancy, one affects radical growth while the other affects shoot growth.
  - Drying; some plants including a number of grasses and those from seasonally arid regions need a period of drying before they will germinate, the seeds are released but need to have a lower moister content before germination can begin. If the seeds remain moist after dispersal, germination can be delayed for many months or even years. Many herbaceous plants from temperate climate zones have physiological dormancy that disappears with drying of the seeds. Other species will germinate after dispersal only under very narrow temperature ranges, but as the seeds dry they are able to germinate over a wider temperature range.
  - Photodormancy or light sensitivity affects germination of some seeds. These photoblastic seeds need a period of darkness or light to germinate. In species with thin seed coats, light may be able to penetrate into the dormant embryo. The presence of light or the absence of light may trigger the germination process, inhibiting germination in some seeds buried too deeply or in others not buried in the soil.

– Thermodormancy is seed sensitivity to heat or cold. Some seeds including cocklebur and amaranth germinate only at high temperatures (30C or 86F) many plants that have seed that germinate in early to mid summer have thermodormancy and germinate only when the soil temperature is warm. Other seeds need cool soils to germinate, while others like celery are inhibited when soil temperatures are too warm. Often thermodormancy requirements disappear as the seed ages or dries.

Combinational dormancy also called double dormancy. Many seeds have more than one type of dormancy, some *Iris* species have both hard impermeable seeds coats and physiological dormancy.

Secondary dormancy is caused by conditions after the seed has been dispersed and occurs in some seeds when none dormant seed is exposed to conditions that are not favorable to germination, very often high temperatures. The mechanisms of secondary dormancy is not yet fully understood but might involve the lose of sensitivity in receptors in the plasma membrane.

Many garden plants have seeds that will germinate readily as soon as they have water and are warm enough, though their wild ancestors may have had dormancy, these cultivated plants lack seed dormancy. After many generations of selective pressure by plant breeders and gardeners dormancy has been selected out.

For annuals, seeds are a way for the species to survive dry or cold seasons. Ephemeral plants are usually annuals that can go from seed to seed in as few as six weeks.

Not all seeds undergo a period of dormancy. Seeds of some mangroves are viviparous, they begin to germinate while still attached to the parent. The large, heavy root allows the seed to penetrate into the ground when it falls.

Chapter 2

Seed Structure

A seed contains the embryo from which a new plant will grow under proper conditions. Seeds also usually contain a supply of stored energy and is wrapped in the seed coat or testa. Seeds are very diverse in size. The dust-like orchid seeds are the smallest with about one million seeds per gram. Embryotic seeds have immature embryos and no significant energy reserves. They are myco-heterotrophs, depending on mycorrhizal fungi for nutrition during germination. At over 20 kg, the largest seed is the coco de mer.

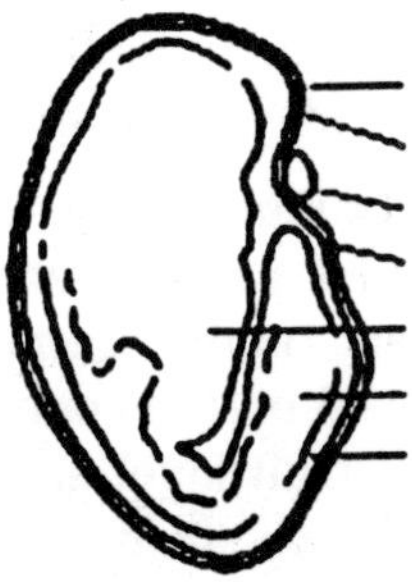

Fig. Alfalfa seed structure

The embryo has one cotyledon or seed leaf in monocotyledons, two cotyledons in almost all dicotyledons and two or more in gymnosperms. The radicle is the embryonic root. The plumule is the embryonic shoot. The embryonic stem above the point of attachment of the cotyledon(s) is the epicotyl. The embryonic stem below the point of attachment is the hypocotyl.

In angiosperms, the stored food begins as a tissue called the endosperm, which is derived from the parent plant via double fertilization. The usually triploid endosperm is rich in oil or starch and protein. In gymnosperms, such as conifers, the food storage tissue is part of the female gametophyte, a haploid tissue. In some species, the embryo is embedded in the endosperm or female gametophyte, which the seedling will use upon germination.

In others, the endosperm is absorbed by the embryo as the latter grows within the developing seed, and the cotyledons of the embryo become filled with this stored food. At maturity, seeds of these species have no endosperm and are termed exalbuminous seeds. Some exalbuminous seeds are bean, pea, oak, walnut, squash, sunflower, and radish. Seeds with an endosperm at maturity are termed albuminous seeds. Most monocots (e.g. grasses and palms) and many dicots (e.g. brazil nut and castor bean) have albuminous seeds. All gymnosperm seeds are albuminous. The seed coat develops from the tissue, the integument, originally surrounding the ovule.

The seed coat in the mature seed can be a paper-thin layer (e.g. peanut) or something more substantial (e.g. thick and hard in honey locust and coconut). The seed coat helps protect the embryo from mechanical injury and from drying out.

The seeds of angiosperms are contained in a hard or fleshy (or with layers of both) structure called a fruit. Gymnosperm seeds begin their development "naked" on the bracts of cones, although the seeds do become covered by the cone scales as they develop. An example of a hard fruit layer surrounding the actual seed is that of the so-called *stone* fruits (such as the peach).

Some seeds have an appendage on the seed coat such an aril (as in yew and nutmeg) or an elaiosome (as in Corydalis) or hairs (as in cotton). The hilum is the scar on the seed coat where the seed was attached to the ovary wall by the funiculus.

In order for the seed coat to split, the embryo must imbibe (soak up water), which causes it to swell, splitting the seed coat. However, the nature of the seed coat determines how rapidly water can penetrate and subsequently initiate germination. For seeds with a very thick coat, scarification of the seed coat may be necessary before water can reach the embryo.

Examples of scarification include: gnawing by animals, freezing and thawing, battering on rocks in a stream bed, or passing through an animal's digestive tract. In the latter case, the seed coat protects the seed from digestion, while perhaps weakening the seed coat such that the embryo is ready to sprout when it gets deposited (along with a bit of fertilizer) far from the parent plant. In species with thin seed coats, light may be able to penetrate into the dormant embryo.

The presence of light or the absence of light may trigger the germination process, inhibiting germination in some seeds buried too deeply or in others not buried in the soil. Abscisic acid is usually the growth inhibitor in seeds.

A fertilized seed contains the embryo from which a new plant will grow under proper conditions. It also contains a supply of stored food and is wrapped in the seed coat or testa. The stored food begins as a tissue called endosperm derived from the parent plant. Endosperm becomes rich in oil or starch, and protein. In some species, the embryo is imbedded in the endosperm, which the seedling will use upon germination. In others, the endosperm is absorbed by the embryo as the latter grows within the developing seed, and the ...

A fertilized seed contains the embryo from which a new plant will grow under proper conditions. It also contains a supply of stored food and is wrapped in the seed coat or *testa*. The stored food begins as a tissue called *endosperm* derived from the parent plant. Endosperm becomes rich in oil or starch, and protein.

In some species, the embryo is imbedded in the *endosperm*, which the seedling will use upon germination. In others, the endosperm is absorbed by the embryo as the latter grows within the developing seed, and the cotyledons of the embryo become filled with this stored food. At maturity, seeds of these species have no endosperm. Some common plant seeds that lack an endosperm are bean, pea, oak, walnut, squash, sunflower, and radish. Plant seeds with an endosperm include all conifers and most monocotyledons (e.g. grasses and palms), and also e.g. brazil nut, castor bean.

The seed coat develops from tissues (called *integument*) originally surrounding the ovule. The seed coat in the mature seed can be a paper thin layer or something more substantial. The seed coat helps protect the embryo from mechanical injury and from drying out. In order for the seed coat to split, the embryo must imbibe (soak up water) which causes it to swell, splitting the seed coat.

However, the nature of the seed coat determines how rapidly water can penetrate and subsequently initiate germination. For seeds with a very thick coat, scarification of the seed coat may be necessary before water can reach the embryo. Examples of scarification include: gnawing by animals, freezing and thawing, battering on rocks in a stream bed, or passing through an animal's digestive tract.

In the latter case, the seed coat protects the seed from digestion, while perhaps weakening the seed coat such that the embryo is ready to sprout when it gets deposited (along with a bit of fertilizer) far from the parent plant. In species with thin seed coats, light may be able to penetrate into the dormant embryo. The presence of light or the absence of light may trigger the germination process, inhibiting germination in some seeds buried too deeply or in others not buried in the soil. Abscisic acid is usually the growth inhibitor in seeds.

The seeds of angiosperms are contained in a hard or fleshy (or with layers of both) structure called a fruit. Gymnosperm seeds begin their development "naked" on the bracts of cones, although the seeds do become covered by the cone scales as they develop. An example of a hard fruit layer surrounding the actual seed is that of the so-called *stone* fruits (such as the peach).

Alfalfa is a dicotyledonous plant. That is, the seed is composed of two embryonic or seed leaves called cotyledons. In addition, the seed contains the primary root or radicle, the shoot growing point or epicotyl located above the cotyledons, and the endosperm or food storage area. Externally the seed has several visible structures – the hilum or point of attachment in the seed pod, the lens or a weak point in the seed coat, and the micropyle, which is a remnant of the tiny opening the pollen tube grew through during the process of fertilizing the female ovary.

ALFALFA SEED STRUCTURE

Germination is the resumption of growth by the embryo within the seed leading to the development of a new plant. The germination process is complete as soon as the radicle ruptures the seed coat.

The process is influenced by available soil moisture, soil temperature, and the nature of the soil surrounding the seed (i.e., salty vs. normal soil) or residual herbicides that may be present in the soil.

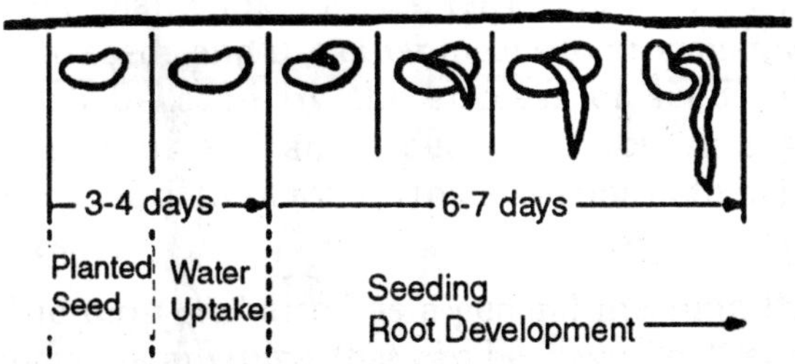

Fig. Germination and seedling root growth

ALFALFA GERMINATION PROCESS

Seedling growth is the developmental period of the young plant from the time germination is completed until it can manufacture enough food through photosynthesis to sustain growth. The seedling root is the first structure to emerge from the seed during germination. It penetrates the soil very rapidly, forming a slender, usually unbranched taproot, which may penetrate 5 to 6 feet into the soil during the first growing season.

Once the seedling root is anchored firmly in the soil, the seedling axis below the cotyledons elongates in an arch (hypocotyl arch) pulling the cotyledons upward to the soil surface. Seed germination and seedling emergence occur in about three to seven days. Fewer days may be required for seedling emergence under ideal soil moisture and temperature conditions. As the hypocotyl arch emerges from the soil, growth stops on the side exposed to light and continues on the underside until the seedling is in an upright position.

This raises the cotyledons above the soil surface where they expand. The growing point (epicotyl) of the seedling is now exposed. The first true leaf is unifoliolate and emerges from a bud at the first stem node above the cotyledons. Seedling growth is complete. Under good growing conditions, the seedling is developed fully 10 to 15 days after planting.

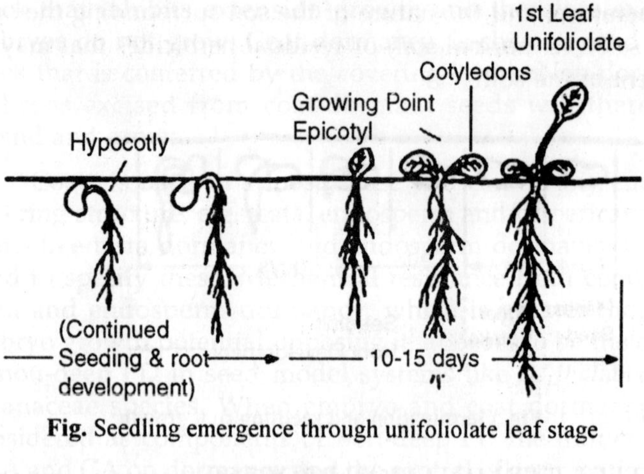

Fig. Seedling emergence through unifoliolate leaf stage

ALFALFA SEEDLING GROWTH

A fully developed alfalfa seedling does not assure plant establishment. It must continue to develop deeper roots and grow more leaves to survive and become an established stand. Vegetative growth continues through cell division and expansion in the epicotyl or growing point of the young plant.

The second leaf of the alfalfa plant is usually trifoliolate (three leaflets) and originates from the second primary stem node. All subsequent leaves are trifoliolates except in new multifoliolate varieties that have 5, 7, or 9 leaflets per leaf. The unifoliolate leaf and all subsequent trifoliolate/multifoliolate leaves are useful characters to distinguish alfalfa from weed seedlings in stand evaluation.

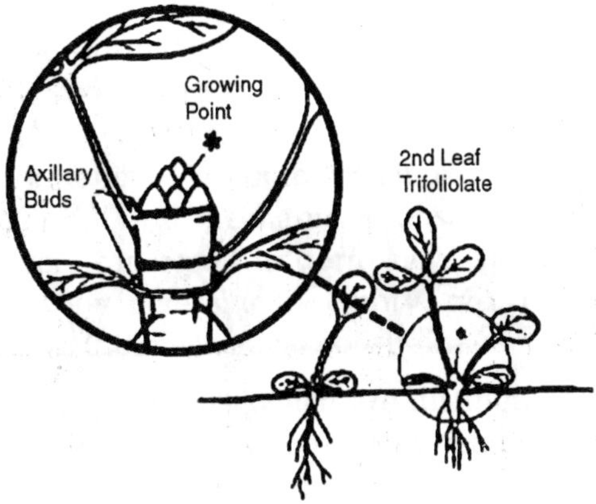

Fig. Unifoliolate through fourth trifoliolate leaf stage

Axillary buds develop in the axils of all leaves. After three or more trifoliolate/multifoliolate leaves have appeared on the primary stem, new secondary stem growth may occur from any of the axillary buds, but frequently only one axillary bud stem (usually the unifoliolate) develops early, especially if seedlings are shaded in a companion crop or under competition from weed growth or other alfalfa plants.

ALFALFA FIRST TRIFOLIOLATE LEAF AND BUDS

The primary and secondary stem(s) of the young plant increase in length by cell division and internode elongation from the first stem node upwards. The second and subsequent leaves are trifoliolate/multifoliolate and develop alternately at each stem node as growth continues.

Once the first true trifoliolate leaf develops, further growth and development is best described by the number of trifoliolate/multifoliolate leaves that develop on the main shoot as the plant continues to grow. Any axillary buds in axils of leaves can develop new stem tissue and generally do in less competitive conditions.

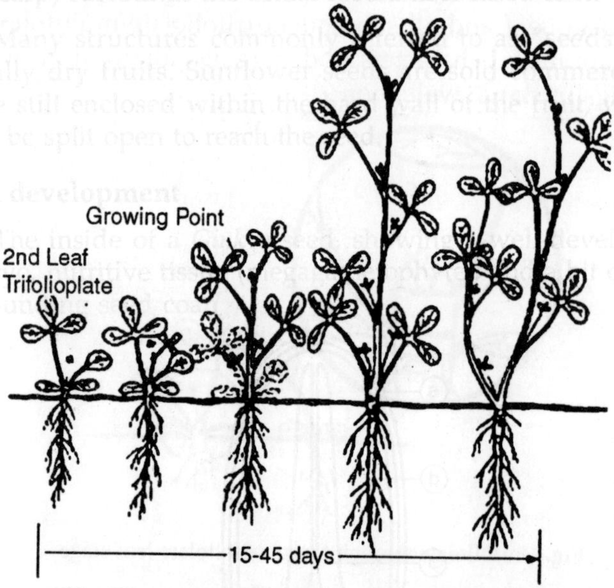

Fig. First-year vegetative growth and development

ALFALFA FIRST-YEAR VEGETATIVE DEVELOPMENT

Growth and development of new shoots from axillary buds gives the young plant a branched appearance, especially if light is adequate and the stand is not too thick. Vigorously growing alfalfa plants quite often have three and sometimes four secondary shoots in addition to the primary or central stem, which forms the characteristic first-year crown.

The branches from the unifoliolate leaf bud appears first, followed by branches from the cotyledonary and first trifoliolate leaf buds. Fewer branches may form the crown with late seedings, under severe companion crop competition, or with high seeding rates. Under these conditions the cotyledonary, unifoliolate, and first trifoliolate leaf buds may remain dormant or may give rise to a branch rhizome or underground stem in the fall.

Fig. Crown development showing contractile growth

The young alfalfa plant, at about 8 to 10 weeks of age, undergoes a growth phase known as contractile growth. This process in alfalfa and sweetclover involves a change in the shape of cells in the hypocotyl or seedling axis below the cotyledons and upper portion of the primary root from long and narrow to short and wide as a result of carbohydrate or food storage. This shift pulls the lower stem nodes beneath the soil surface.

Most winter hardy alfalfa varieties have several nodes pulled well below the soil surface (1 to 3 inches) in the seeding year. Contractile growth greatly aids winter survival of alfalfa by providing soil insulation for the perennial over-wintering crown structures.

CROWN DEVELOPMENT

The lower internodes or nodes of the primary or secondary stems may produce adventitious or crown buds. Crown buds or dominant axillary buds produce the vegetative regrowth following harvest the seedling year, or these buds may give rise to branch rhizomes in the fall. The branch rhizome is the structure from which new growth will initiate in the spring unless winter injury occurs.

The branch rhizome must be developed adequately in the fall by the first killing frost or the seeding will fail, even under the most ideal growing conditions. In most years, alfalfa seeded later than about August 15 under good moisture (irrigation) conditions in North Dakota will not have adequate time to initiate branch rhizomes before the first killing frost. A dryland seeding after the first week of August is not recommended because of the increased risk of stand loss.

A crown of a perennial can be described best as the young overwintering stem tissue. The crown in the fall of the first year may be as small as the lower portion of the main stem and dormant cotyledonary node buds, or as large as the lower portion of the main stem (i.e., colytedonary, unifoliolate, and first trifoliolate leaf nodes), the secondary stems from these nodes, and the branch rhizomes that develop.

Spring growth will occur from the rudimentary leaves on the branch rhizome if winter injury hasn't "burnt" them off. If winter injury has occurred, spring growth will initiate from adventitious buds on the branch rhizome or from a dormant crown bud. But, spring growth is much slower in this case. All subsequent growth in the first-harvest year occurs from adventitious or dormant crown buds.

The crown of the alfalfa plant increases in size during the second year. Branch rhizomes develop on last year's branch rhizome. These grow outward and upward increasing the circumference of the plant. After two to four years, the typical multibranched crown of alfalfa develops.

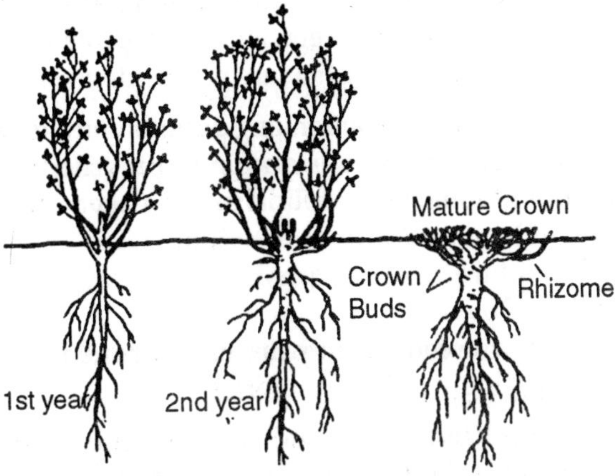

Fig. Fully developed alfalfa crown

ALFALFA WINTERING STRUCTURES

The crown and associated root tissue are the storage center for carbohydrates (food) produced through the process of photosynthesis. The carbohydrates are used to develop cold resistance or winter hardiness, for plant respiration during winter, and to provide the energy source for the initiation of the new growth in the spring and following each harvest. First-year management influences subsequent growth. Alfalfa management is discussed in R-571, Alfalfa Management in North Dakota.

Over the next several lectures we will follow the early growth of young trees and other vascular plants from seed through germination and growth in the first year. We will draw examples from many species, but will illustrate some phases of growth by contrasting two tree species native to British Columbia: Douglas-fir representing gymnosperms, and red alder representing angiosperms.

Evolution of Seeds

The evolution of seeds was a critical factor in facilitating the proliferation of land plants into a wide range of environments. Seeds protect and nourish embryos, and embryos can survive a wide range of environmental conditions. Seeds can also persist in a dormant state, enabling plant species to survive and regenerate following catastrophic events or periods of weather that are not conducive for growth.

Seeds can remain viable for long periods of time, centuries, even millenia. An extreme example is provided by frozen 10,000 year old lupine seeds found in the Yukon. These seeds germinated within 48 hours of thawing!

Gymnosperms

Gymnosperm seeds are, by definition, naked, that is they lack a surrounding fruit. The outermost

layer of the seed is the seed coat. It originates from the mother tree and is diploid. The seed coat has three layers: the outer layer; the thicker, tough 'stony' or middle layer; and the inner layer. The colour of the seed coat, which can be quite variable, is due to tannins that accumulate during seed development. Some species of conifers, including *Abies* spp. (true firs), *Tsuga* spp. (hemlocks) and *Thuja plicata* (western redcedar) have resin vesicles in the middle or outer layers of the seed coat.

These resin vesicles may play a role in:

1. Seed coat dormancy,
2. Protecting the embryo from excessive dehydration,
3. Deterring seed herbivory.

These vesicles can also create problems for the extraction and cleaning of conifer seed. Many conifer seeds have a seed wing to aid in dispersal.

Immediately inside the cell wall is the nucellus, a papery layer surrounding the megaspore cell wall. Inside the megaspore cell wall is the megagametophyte (the female gametophyte), the haploid nutritional tissue found in gymnosperm seeds. Early in the development of cones, megagametophytes produce egg cells which are fertilized by male gametes to produce zygotes. Zygotes develop into embryos. The megagametophyte then plays its second functional role, surrounding the embryo, protecting and nourishing it.

The megagametophyte in Douglas-fir is 60% lipids, 16% proteins, and 2% sugars, making it a high-energy and nutrient tissue both for the embryos it contains and for a plethora of seed predators including small mammals, birds and insects. In a dry seed (<10% moisture), the megagametophyte is creamy yellow and does not entirely fill the seed coat. When a seed is imbibed (soaked in water until the moisture content is 30-40%), the megagametophyte is white and shiny.

The embryo is found in the corrosion cavity, a pit in the centre of the megagametophyte that is fully filled by the embryo in mature seeds. It consists of the cotyledons, shoot apical meristem, root apical meristem, root cap and suspensor. The cotyledons and shoot apical meristem point towards the wider chalazal end of the seed.

The radicle (embryonic root including the apical meristem and root cap) and suspensor are at the more pointed micropylar end of the seed towards the micropyle (the opening in the integument through which the pollen entered the ovule). The number of cotyledons varies from 2 in western redcedar to 12 or more in *Pinus ponderosa* (ponderosa pine).

The suspensor is found at the base of the root cap and plays a role early in embryo development by pushing the embryo into the megagametophyte.

Angiosperms

Angiosperm seeds are found inside fruit. The many types of fruit will be discussed in detail in lab. Angiosperm seeds generally vary more in structure than conifer seeds, but share some features with conifer seeds and have different parts playing the same functional roles for others.

Like conifer seeds, angiosperm seeds are surrounded by a seed coat protecting the embryo. Inside the seed coat (which is usually formed by the integument, but is sometimes fused with the fruit), nutritive reserves for the mature embryo may be stored either in endosperm or in large cotyledons.

Endosperm is formed through the fertilization of the two polar nuclei by a male gamete, producing triploid cells, at the same time that the egg cell is fertilized by a second male gamete (more on this after the Christmas break). All angiosperm seed have endosperm early in embryo development. In mature seeds of some species, there may be a substantial amount of triploid endosperm surrounding the embryo (e.g., *Arbutus menziesii* (arbutus)).

Commonly in Eudicots, however, embryos largely absorb the endosperm and the nutritional reserves are stored in the two (of course!) large cotyledons that fill the seed cavity (e.g., *Alnus* spp. (alders), *Acer* spp. (maples), and *Quercus* spp. (oaks)). Some angiosperm seeds already contain a primordial shoot with leaves, called a plumule, above (distal to) the cotyledons prior to germination.

The seeds of monocots have several noteworthy features. They typically have endosperm at maturity, and the single cotyledon often plays a role in absorbing food digested from the endosperm. Grass embryos have a large cotyledon called a scutellum attached to one side, as well as a plumule and a radicle. The plumule is protected by a structure called the coleoptile, while the radicle is protected by the coleorhiza.

Angiosperm seed also have suspensors, but these have a different function than those of conifers. In angiosperms, suspensors are involved in supporting the early development of the embryo both through providing nutrition and growth regulators. The suspensor cells die in angiosperms relatively early in embryo development.

SEED DISPERSAL

Conifers

The dispersal of winged seeds depends on seed mass, wing size and wind speed. Douglas-fir seeds have a mass of 9 to 17 mg, are 1 to 1.5 cm in length including wings, and can disperse up to 3 or 4 tree heights from the mother tree. Periodic weather events may result in rare dispersals up to a km in distance.

Most BC conifers shed seed August through October. Cone tissues die, cones dry out, cone scales flex and release seed. Dry conditions promote the flexing of cone scales and opening of cones, so cycles of wet and dry weather may produce repeated cycles of seed release. Seeds that land on snow may blow considerable distances along the frozen surface in winter.

Not all conifers have cones that open on maturity. The true firs (*Abies* spp.) bear upright rather than pendant cones, and have deciduous cone scales. When these cones are mature, the cone scales fall off the central cone rachis, releasing the seed. Whitebark pine (*Pinus albicaulis*) cones do not open and must be mechanically torn apart to release seeds. This is done by Clark's nutcrackers, which then disperse store seeds for later consumption (and regeneration of the pine).

Angiosperms

Dispersal of seeds in angiosperms can be via wind, through animal ingestion or through animal

transport to food caches or as 'hitchhikers' (e.g., burrs). The acorns of oaks, and the many fleshy fruits and tasty nuts produced by angiosperms, provide examples of angiosperms dispersed by animals.

Red alder is a wind-dispersed species with single seeds found in small (1.5 million per kg), winged nutlets. The fruit are shed in large numbers in mid to late fall, and can be wind-dispersed in air or across snow. Red alder is a good example of a pioneer species, capable of colonizing bare ground. Pioneer species typically have small seed produced in large numbers that disperse widely via wind.

Species of poplars and aspen (*Populus* spp.) are also pioneers, and produce large quantities of tiny seed (8 million per kg for trembling aspen). Poplar seed contain very small nutritional reserves and thus are quite short-lived. Fireweed (*Epilobium angustifolium*) is a good example of an herbaceous pioneer with seed production and dispersal similar to poplars.

SEED DORMANCY

Seed of most temperate and boreal species is dormant when it disperses, that is it won't germinate even if conditions of temperature, moisture and light are favorable. Why? To avoid active growth until winter has passed.

An actively growing plant cannot be adequately frost hardy. Dormant seeds cease growth and maintain a very low metabolic rate until specific environmental signals are received or other requirements have been met.

Embryo Dormancy

Dormancy in many species is under the physiological control of the embryo. Embryo dormancy results from the levels of plant growth regulators (plant hormones). Dormant embryos have growth-inhibitors present and lack growth promoters. Breaking embryo dormancy is dependent on decreasing growth inhibitor concentrations and increasing growth promoters above some threshold.

Embryo dormancy in many species can be overcome through exposure to low temperatures to meet the chilling requirement of the seed. To accomplish this, seeds are imbibed (soaked in water), then placed in a fridge, usually on damp paper or moss. This is called stratification.

In Douglas-fir, holding seeds at 4°C for 4 weeks will meet the chilling requirement. Many species require a longer stratification period – up to 100 days. Seeds of some species with embryo dormancy may germinate eventually without chilling after being soaked in water, but will germinate more slowly than stratified seed, and the percent of seeds that germinate will likely be lower.

Embryo dormancy can also be overcome in some cases by artificially applying growth regulators. Some species require alternating day and night temperatures, or cycles of wetting and drying following low temperatures. Some species have both embryo dormancy and other types of seed dormancy.

Seed Coat Dormancy

Some desert and tropical species have chemical growth inhibitors in the seed coat that must be

leached out by water before seed will germinate, increasing the probability that seed will germinate during adequately moist periods. Other species require a heat treatment to germinate. Germination of buckbrush (*Ceonothus* spp.) is promoted by pouring boiling water over seeds.

Other Types of Seed Dormancy

Embryo immaturity – Seeds of some species are immature at the time fruit or seed are shed and require 'after-ripening'. The seed of whitebark pine are typically immature when harvested by Clark's nutcrackers and continue to mature after caching.

Seed coat impermeability – Seed coats can be impermeable to water or oxygen, which physically imposes dormancy through preventing growth. This is common for leguminous species, among others. These seed may need scarification (scratching or abrading of the seed coat, either in nature or artificially with sandpaper), infection of seed coat by fungi or bacteria, or treatment with concentrated sulphuric acid (simulating passing the seed through the digestive tract of an animal).

Light requirement – Some species (e.g, some spruce and pine) require exposure to light to germinate. Other species require exposure to light of a specific quality. Red light promotes germination in these species, while far red light inhibits it. It is the ratio of far red to red light that is critical. As light passes through vegetation, this ratio increases, and germination in light-sensitive species will decrease. This means that seeds will be less likely to germinate when they are below competing vegetation. Once the competing vegetation is gone, the amount of red to far red light will increase, and the seed will germinate.

Serotiny – Most interior lodgepole pine (*Pinus contorta* ssp. *latifolia*) has cones that are serotinous. The cone scales are 'glued' closed by resin that has a melting point around 40°C. These cones must be exposed to either a fire or to particularly warm temperatures (e.g., on a dark surface in the full sun) to open. The cones of juvenile interior lodgepole pine, and Shore pine (*P. contorta* ssp. *contorta*) do not have this adaptation – why?

Species Without Seed Dormancy

Red alder seeds lack dormancy, but since they are shed late in the fall, environmental conditions are not favorable for germination until spring. Other examples of seeds with no dormancy include sugar maple and willows, which are viable for only a few days following dispersal.

Many agricultural crop species have no seed dormancy, but their wild progenitors do – humans have selected against seed dormancy in these species as it complicates farming.

GERMINATION

Germination is the resumption of growth by embryos when exposed to favorable conditions. It requires several steps:

1) seeds imbibe and swell; the seed coat ruptures;

2) cell division occurs in the embryo and the radicle elongates and extends out of the seed;

3) further elongation of the radicle pushes either the cotyledons or just the epicotyl hormones produced in the meristems and possibly elsewhere initiate enzymatic activity, hydrolysis of stored nutrients, and movement of these nutrients to regions of growth.

The process of germination is complete when enough photosynthetic tissue is exposed to support further growth.

The radicle develops into the tap root of the seedling (except in the case of species with fibrous root systems). It is positively gravitropic (grows in the direction of gravitational pull) while the shoot is negatively gravitropic (grows against gravity). The root cap consists of loose, slimy cells that protect the root apical meristem as the root elongates through the soil. The radicle develops root hairs for absorption of water and dissolved nutrients soon after germination.

In species with epigeous germination (e.g., conifers, red alder, maples), the cotyledons are pushed out of the ground, turn green and become photosynthetically active. In species with hypogeous germination (e.g., oaks and other species with large seeds), the cotyledons remain in the ground and only the epicotyl emerges. The energy reserves in the cotyledons supply the energy for early seedling growth, but the cotyledons never become photosynthetically active in this case. Rather, the first true leaves are the first photosynthetic organs with hypogeous germination.

Symbiotic relationships between seedling roots and other organisms are often initiated soon after germination. Lateral roots can become infected by mycorrhizal fungi soon after germination. In addition to becoming mycorhizal, red alder germinants becomes symbiotic with *Frankia* bacteria. Nodules form on the roots, and in these nodules N_2 gas is fixed into NH_4^+, which can then be converted into amino acids. This allows red alder to 'fertilize itself'. Up to 320 kg N can be fixed per ha per year! Other species that fix nitrogen include *Ceonothus* spp. (deer brush) and *Shepherdia canadensis* (soopallalie).

SEED STRUCTURE AND DEVELOPMENT

Following the fertilizations in the embryo sac, the zygote divides repeatedly by mitosis and differentiates into an embryo. The endosperm nucleus also divides by mitosis and forms the endosperm tissue, which provides food for the developing embryo.

The early embryo is linear with apical meristems on either end and one or two seed leaves or cotyledons. The axis below the cotyledons is called the hypocotyl, at the tip of which is the radicle that gives rise to the primary root of the seedling.

The axis above the attachment of the cotyledons is the epicotyl, which also ends in an apical meristem. In some seeds, the first foliage leaves are formed in the seed.

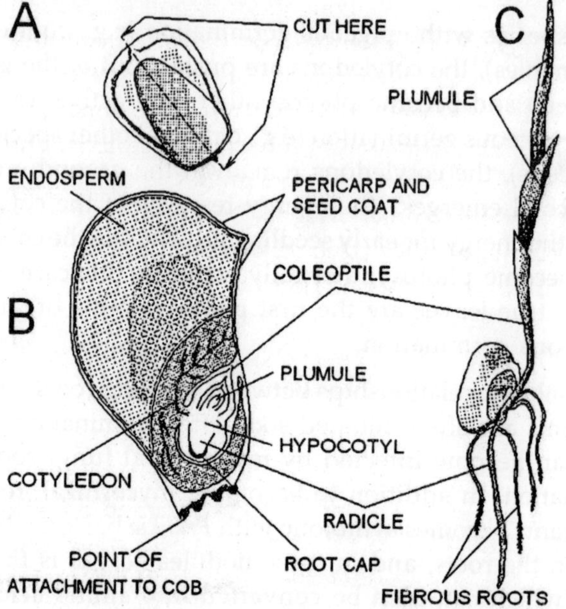

Fig. Seed Structure Monocot.

The area above the cotyledons is thus a miniature shoot and is called the plumule. In some taxa, food for the embryo remains within endosperm tissue and the cotyledons serve as organs of absorption. In others, food moves directly into the embryo and is stored within the cotyledons, leaving only a minuscule endosperm. In still another variation, the nucellus (the megasporangium wall) enlarges and becomes a storage tissue called perisperm. The integuments harden into the seed coat as the embryo matures.

The scar left on the seed coat by the separation of the funiculus from the integuments is called the hilum. Often the micropyle remains visible near the hilum.

Angiosperms traditionally have been separated into two major categories on the basis of the number of cotyledons they possess: monocots (mono = one; cotyledons = seed leaves) and dicots (di = two; cotyledons = seed leaves). Monocots are the grasses, sedges, lilies, and their relatives, while dicots constitute the rest of the flowering plants.

This artificial separation currently is being replaced by a more natural classification in which the *monocots* are retained as a natural group, but the dicots are separated into the eudicots (eu = true, dicots) and the magnoliids. The latter is a small group of very primitive angiosperms ancestral to both monocots and eudicots and has both woody representatives—woody magnoiiids (trees like *Magnolias*, tulip trees, and laurels)—and paleoherbs (herbaceous plants such as members of the pepper and water lily families). The eudicots constitute about 97 percent of the angiosperms, while the magnoliids make up the other 3 percent.

In monocot seeds the single cotyledon usually digests and absorbs food from the endosperm and translocates it to the embryo. In grasses, the cotyledon is called the scutellum. In grasses, also, there is a protective sheath called the coleoptile over the plumule and another, the coleorhiza, surrounding the radicle.

SEED GERMINATION

Internal signals shut off growth of the embryo at a certain size and the seed goes into a period of dormancy. In some plants, dormancy lasts only as long as it takes the seed to be dispersed from the ovary. In others, dormancy may last for long periods until either external or internal signals (or a combination of both) initiate further growth. Environmental factors of chief importance to initiate growth are water, light, and temperature.

SEEDLING GROWTH

When an embryo resumes growth, stored food provides the energy for seedling development—the roots first, followed by elongation of the photosynthetic shoots.

Eudicot Development

Aboveground growth in eudicots takes one of two general patterns: epigeous or hypogeous. In epigeous growth, the hypocotyl elongates, pulling the plumule and cotyledons above ground; in hypogeous growth, the cotyledons remain below ground because the epicotyl grows faster than the hypocotyl and pulls the plumule erect.

Monocot Development

Monocots develop with two different general patterns: one for the grasses, one for the rest of the group. In most of the monocots (but not grasses), after the radicle has pushed out of the seed coat, the first shoot structure to emerge is the cotyledon, which arches upward with the remainder of the endosperm and the seed coat still attached. It elongates above ground and is photosynthetic until the true leaves develop.

In the grasses, the sheaths around both the shoot and the root tip must be penetrated by the roots and the shoots. The root sheath, the coleorhiza, grows faster than the radicle for a short time, but when it stops growing, the radicle emerges and forms an anchoring primary root. The shoot sheath, the coleoptile, moves upward to the soil surface through elongation of the first internode of the stem (called the *mesocotyl*), and when it reaches the surface it stops growing.

The plumule then pushes through into the air. At about the same time, buds of adventitious roots begin to grow and, by the time the seedling is erect with a few true leaves, it already has adventitious roots growing downward from its first node. The primary root system is short lived and dies soon after its establishment and the adventitious root system becomes the principal absorbing and anchoring system for the new grass plant.

GERMINATION

Germination is the process whereby growth emerges from a period of dormancy. The most common example of germination is the sprouting of a seedling from a seed of an angiosperm or gymnosperm.

Fig. Sunflower seedlings, just three days after germination

However, the growth of a sporeling from a spore, the growth of hyphae from fungal spores, is also germination. In a more general sense, germination can imply anything expanding into greater being from a small existence or germ.

Pollen Germination

Another germination event during the life cycle of gymnosperms and flowering plants is the germination of a pollen grain after pollination. Like seeds, pollen grains are severely dehydrated before being released to facilitate their dispersal from one plant to another. They consist of a protective coat containing several cells (up to 8 in gymnosperms, 2-3 in flowering plants). One of these cells is a tube cell.

Once the pollen grain lands on the stigma of a receptive flower (or a female cone in gymnosperms), it takes up water and germinates. Pollen germination is facilitated by hydration on the stigma, as well as the structure and physiology of the stigma and style. Pollen can also be induced to germinate *in vitro* (in a petri dish or test tube).

During germination, the tube cell elongates into a pollen tube. In the flower, the pollen tube then grows towards the ovule where it discharges the sperm produced in the pollen grain for fertilization. The germinated pollen grain with its two sperm cells is the mature male microgametophyte of these plants.

Self-incompatibility

Since most plants carry both male and female reproductive organs in their flowers, there is a high risk for self-pollination and thus inbreeding. Some plants use the control of pollen germination as a way to prevent this selfing.

Germination and growth of the pollen tube involve molecular signaling between stigma and pollen. In self-incompatibility in plants, the stigma of certain plants can molecularly recognize pollen from the same plant and prevents it from germinating.

SPORE GERMINATION

Germination can also refer to the emergence of cells from resting spores and the growth of sporeling hyphae or thalli from spores in fungi, algae, and some plants.

Resting Spores

In resting spores, germination involves cracking the thick cell wall of the dormant spore. In zygomycetes the thick-walled zygosporangium cracks open and the zygospore inside gives rise to the emerging sporangiophore. In slime molds, germination refers to the emergence of amoeboid cells from the hardened spore. After cracking the spore coat, further development involves cell division, but not necessarily the development of a multicellular organism.

Zoospores

In motile zoospores, germination frequently means a lack of motility and changes in cell shape, which allow the organism to become sessile.

Ferns and Mosses

In plants such as bryophytes, ferns, and a few others, spores germinate into independent gametophytes. In the bryophytes (e.g. mosses and liverworts), spores germinate into protonemata, similar to fungal hyphae, from which the gametophyte grows. In ferns, the gametophytes are small, heart-shaped prothalli that can often be found underneath a spore-shedding adult plant.Seedling

A seedling is a young plant sporophyte developing out of a plant embryo from a seed. Seedling development starts with germination of the seed. A typical young seedling consists of three main parts: the radicle (embryonic root), the hypocotyl (embryonic shoot), and the cotyledons (seed leaves).

The two classes of flowering plants are distinguished by their numbers of seed leaves: Monocotyledons (monocots) have one blade-shaped cotyledon, whereas dicotyledons (dicots) have two round cotyledons. Gymnosperms are more varied. Pine seedlings have up to eight cotyledons. The seedlings of some flowering plants have no cotyledons at all. These are said to be acotyledons.

Germination and Early Seedling Development

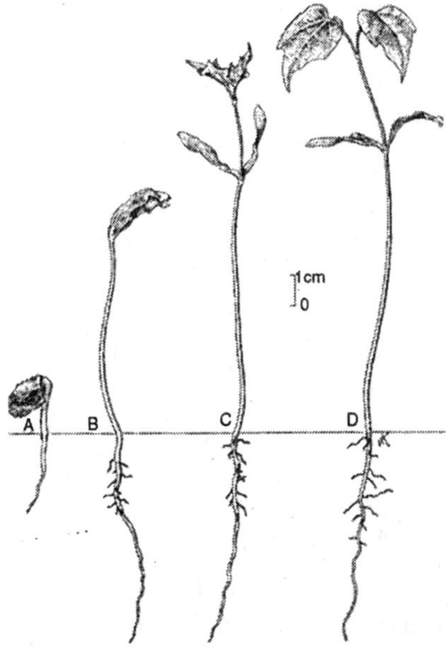

Fig. Development of an angiosperm (maple) seedling

During germination, the young plant emerges from its protective seed coat with its radicle first, followed by the cotyledons. The radicle orients towards gravity, while the hypocotyl orients away from gravity and elongates through cell expansion to push the cotyledons out of the ground.

PHOTOMORPHOGENESIS AND SKOTOMORPHOGENESIS

Dicot seedlings grown in the light develop short hypocotyls and open cotyledons exposing the epicotyl. This is also referred to as photomorphogenesis. In contrast, seedlings grown in the dark develop long hypocotyls and their cotyledons remain closed around the epicotyl in an *apical hook.*

This is referred to as skotomorphogenesis or etiolation. Etiolated seedlings are yellowish in colour as chlorophyll synthesis and chloroplast development depend on light. They will open their cotyledons and turn green when treated with light.

In a natural situation, seedling development starts with skotomorphogenesis while the seedling is growing through the soil and attempting to reach the light as fast as possible. During this phase, the cotyledons are tightly closed and form the *apical hook* to protect the shoot apical meristem from damage while pushing through the soil. In many plants, the seed coat still covers the cotyledons for extra protection.

Upon breaking the surface and reaching the light, the seedling's developmental programme is switched to photomorphogenesis. The cotyledons open upon contact with light (splitting the seed coat open, if still present) and become green, forming the first photosynthetic organs of the young plant. Until this stage, the seedling lives off the energy reserves stored in the seed. The opening of the cotyledons exposes the shoot apical meristem and the *plumule* consisting of the first *true leaves* of the young plant.

The seedlings sense light through the light receptors phytochrome (red and far-red light) and cryptochrome (blue light). Mutations in these photo receptors and their signal transduction components lead to seedling development that is at odds with light conditions, seedlings that show photomorphogenesis when grown in the dark.

Seedling Growth and Maturation

Once the seedling starts to photosynthesize, it is no longer dependent on the seed's energy reserves. The apical meristems start growing and give rise to the root and shoot. The first "true" leaves expand and can often be distinguished from the round cotyledons through their species-dependent distinct shapes. While the plant is growing and developing additional leaves, the cotyledons eventually senesce and fall off.

Consumption of Seedlings

Seedlings are commonly eaten as a health food. These seedlings are usually labeled sprouts, but in a botanical sense are actually seedlings. There is controversy in whether seedlings, or sprouts, are really worth eating. These seedlings are used many times in organic foods and are eaten for their concentrations of certain vitamins the seedling naturally contains.

This can be viewed out of proportion as the seedling is usually too small to contain enough vitamins or minerals to be in line with the amount they are claimed to obtain.

SPROUTING

Sprouting is the practice of soaking, draining and then rinsing seeds at regular intervals until they germinate, or sprout. This can be a semi - automated or fully automated process when done on a large scale for commercial use.

Seeds that can be Sprouted

One of the most common sprouts is that of the mung bean (*Vigna radiata*), often sold as 'Chinese Bean Sprouts'; another common sprout is the alfalfa sprout and the barley sprout.

Other seeds that can be sprouted include adzuki bean, almond, amaranth, annatto seed, anise seed, arugula, basil, navy bean, pinto bean, lima bean, broccoli, buckwheat, cabbage, canola seed, caragana, cauliflower, celery, chia seed, chickpeas, chives, cilantro (coriander, dhania), clover, cress, dill, fennel, fenugreek, flax seed, garlic, hemp seed, kale, kamut, kat, leek, green lentils, lupins,pearl millet, mizuna, mustard, oats, onion, black-eyed peas, green peas, pigeon peas, snow peas, peanut, psyllium, pumpkin, quinoa, radish, rye, sesame, soybean, spelt, sunflower, tatsoi, triticale, watercress, and wheat berries.

However, many sprouts are in fact toxic when eaten, like kidney beans. Some sprouts can be cooked to remove the toxin, while others will be toxic either way and should be avoided. So before eating any sprouts, find out if that species is edible as a sprout.

With all seeds, care should be taken that they are intended for sprouting or human consumption rather than sowing. Seeds intended for sowing may be treated with chemical dressings. Several countries, such as New Zealand, also require that some varieties of edible seed be heat-treated, thus making them impossible to sprout.

Many varieties of nuts, such as almonds and peanuts, can also be started in their growth cycle by soaking and sprouting, although because the sprouts are generally still very tiny when eaten, they are usually called "soaks."

Moisture, warmth, and in most cases, indirect sunlight are necessary for sprouting. Some sprouts, such as mung beans, can be grown in the dark. Little time, effort or space is needed to make sprouts.

To sprout seeds, the seeds are moistened, then left at room temperature (between 13 and 21 degrees Celsius) in a sprouting vessel. Many different types of vessels can be used. One type is a simple glass jar with a piece of cloth secured over its rim. 'Tiered' clear plastic sprouters are commercially available, allowing a number of "crops" to be grown simultaneously. By staggering sowings, a constant supply of young sprouts can be ensured. Any vessel used for sprouting must allow water to drain from it, because sprouts that sit in water will rot quickly. The seeds will swell and begin germinating within a day or two.

Sprouts are rinsed as little as twice a day, but possibly three or four times a day in hotter climates, to prevent them from souring. Each seed has its own ideal sprouting time. Depending on which seed is used, after three to five days they will have grown to two or three inches in length and will be suitable for consumption.

If left longer they will begin to develop leaves, and are then known as baby greens. A popular baby green is sunflower after 7-10 days. The growth process of any sprout can be slowed or halted by refrigerating until needed.

Common causes for sprouts to turn out inedible:

- Seeds are allowed to dry out
- Seeds are left in standing water
- Temperature is high or too low
- Insufficient rinsing
- Dirty equipment
- Insufficient air flow
- Contaminated source of water
- Poor rate of germination of seed

These problems are easily solved by an automatic sprouter that mists and drains the sprouts at regular intervals. To control temperature, in the winter a warming blanket can be placed under the sprouter, and in the summer small fans in the lid if it's very hot and humid.

Mung beans can be sprouted either in light or dark conditions. Those sprouted in the dark will be crisper in texture and whiter, as in the case of commercially available Chinese Bean Sprouts, but these have less nutritional content than those grown in partial sunlight.

Growing in full sunlight is not recommended, because it can cause the beans to overheat or dry out. Subjecting the sprouts to pressure, by placing a weight on top of them in their sprouting container, will result in larger, crunchier sprouts similar to those sold in Polish grocery stores.

A very effective way to sprout beans like lentils or adzuki is in colanders. Soak the beans in hot water for about 8 hours then place in the colander. Wash twice a day. The sprouted beans can be eaten raw or cooked.

NUTRITIONAL INFORMATION

Sprouts are rich in digestible energy, bioavailable vitamins, minerals, amino acids, proteins, beneficial enzymes and phytochemicals, as these are necessary for a germinating plant to grow. These nutrients are essential for human health. To clarify the nutritional changes upon germination & sprouting have been summarized below.

Chavan and Kadam concluded that - "The desirable nutritional changes that occur during sprouting are mainly due to the breakdown of complex compounds into a more simple form, transformation into essential constituents, and breakdown of nutritionally undesirable constituents."

"The metabolic activity of resting seeds increases as soon as they are hydrated during soaking. Complex biochemical changes occur during hydration and subsequent sprouting. The reserve chemical constituents, such as protein, starch and lipids, are broken down by enzymes into simple compounds that are used to make new compounds."

"Sprouting grains causes increased activities of hydrolytic enzymes, improvements in the contents of total proteins, fat, certain essential amino acids, total sugars, B-group vitamins, and a decrease in dry matter, starch and anti-nutrients. The increased contents of protein, fat, fibre and total ash are only apparent and attributable to the disappearance of starch.

However, improvements in amino acid composition, B-group vitamins, sugars, protein and starch digestibilities, and decrease in phytates and protease inhibitors are the metabolic effects of the sprouting process."

INCREASES IN PLANT ENZYME CONTENT

Sprouts are a tremendous source of (plant) digestive enzymes. Enzymes act as biological catalysts needed for the complete digestion of protein, carbohydrates & fats. The physiology of vitamins, minerals and trace elements is also dependant on enzyme activity.

"Being eaten whilst extremely young, "alive" and rapidly developing, sprouts have been acclaimed as the "most enzyme-rich food on the planet". Estimates suggest there can be up to 100 times more enzymes in sprouts than in fruit and vegetables, depending on the particular type of enzyme and the variety of seed being sprouted. The period of greatest enzyme activity in sprouts is generally between germination and 7 days of age."

"Grains and legume seeds of all plants contain abundant enzymes. However, while grains and seeds are dry, enzymes are largely inactive, due to enzyme inhibitors, until given moisture to activate germination. It is these inhibitors that enable many seeds to last for years in soil without deteriorating, whilst waiting for moisture.

Enzyme inhibitors in some grains and legume seeds need to be inactivated by heating or other processes, before they can be safely fed. However, heating, cooking and grinding processes can also inactivate certain digestive enzymes within grains and seeds. Fortunately, during germination and sprouting of grains and seeds, many enzyme inhibitors are effectively neutralized, whilst at the same time the activity of beneficial plant digestive enzymes is greatly enhanced."

Increases in Crude Protein content Morgan et al. found that -

"The protein content of sprouts increased from the time of germination. The absorption of nitrates facilitates the metabolism of nitrogenous compounds from carbohydrate reserves, thus increasing crude protein levels."

Increases in Protein Quality Chavan and Kadam stated - "Very complex qualitative changes are reported to occur during soaking and sprouting of seeds. The conversion of storage proteins of cereal grains into albumins and globulins during sprouting may improve the quality of cereal proteins.

Many studies have shown an increase in the content of the amino acid Lysine with sprouting."

"An increase in proteolytic activity during sprouting is desirable for nutritional improvement of cereals because it leads to hydrolysis of prolamins and the liberated amino acids such as glutamic and proline are converted to limiting amino acids such as lysine."

Increases in Crude Fibre content Cuddeford, based on data obtained by Peer and Leeson, stated - "In sprouted barley, crude fibre, a major constituent of cell walls, increases both in percentage and real terms, with the synthesis of structural carbohydrates, such as cellulose and hemicellulose".

INCREASES IN ESSENTIAL FATTY ACIDS

An increase in lipase activity has been reported in barley by MacLeod and White, as cited by Chavan and Kadam. Increased lipolytic activity during germination and sprouting causes hydrolysis of triacylglycerols to glycerol and constituent fatty acids.

Sprouting treatment of cereal grains generally improves their vitamin value, especially the B-group vitamins. Certain vitamins such as á-tocopherol (Vitamin-E) and â-carotene (Vitamin-A precursor) are produced during the growth process.

Sprouts provide a good supply of Vitamins A, E & C plus B complex. Like enzymes, vitamins serve as bioactive catalysts to assist in the digestion and metabolism of feeds and the release of energy. They are also essential for the healing and repair of cells. However, vitamins are very perishable, and in general, the fresher the feeds eaten, the higher the vitamin content.

The vitamin content of some seeds can increase by up to 20 times their original value within several days of sprouting. Mung Bean sprouts have B vitamin increases, compared to the dry seeds, of - B1 up 285%, B2 up 515%, B3 up 256%. Even soaking seeds overnight in water yields greatly increased amounts of B vitamins, as well as Vitamin C. Compared with mature plants, sprouts can yield vitamin contents 30 times higher."

Chelation of Minerals Shipard claims that - "When seeds are sprouted, minerals chelate or merge with protein, in a way that increases their function."

Risks and Antinutritional factors Some legumes can contain toxins or antinutritional factors, which can be reduced by soaking, sprouting and cooking (eg, stir frying). Joy Larkcom advises that to be on the safe side "one shouldn't eat large quantities of raw legume sprouts on a regular basis, no more than about 550g daily". However not all legume sprouts contain these antinutritional factors and many have greatly beneficial properties so it is recommended that the advice of a qualified nutritionist is sought before making any decisions about what to include or eliminate from a diet/ration.

Reduction of Anti-Nutritional Factors Phytic Acid occurs primarily in the seed coats and germ of plant seeds. It forms insoluble or nearly insoluble compounds with minerals including Calcium, Iron, Magnesium and Zinc, such that they cannot be effectively absorbed into the blood. Diets high in phytic acid and poor in these minerals produce mineral deficiency symptoms in experimental animals.

The latter authors state that the sprouting of cereals has been reported to decrease levels of Phytic Acid. Similarly, Shipard states that enzymes of germination and sprouting have the ability to eliminate detrimental substances such as Phytic Acid.

Buckwheat greens contain fagopyrin, a naturally occurring substance in the buckwheat plant. When ingested in sufficient quantity, fagopyrin is known to cause the skin of animals and people to become phototoxic, which is to say hypersensitive to sunlight, particularly if juiced or eaten in large quantities. Due to the growing popularity of sprouts in general, and a widespread ignorance as to the toxic dangers posed by buckwheat greens specifically, many people are today suffering unnecessarily.

SPROUTING AND THE LIVING FOODS DIET

Advocates of a raw food diet promote the use of sprouting as an effective way to increase the nutrient value, and digestibility, of beans, seeds and nuts. Sprouts are in fact the most nutrient dense food on earth– nutrient density refers to the amount of nutrients per calorie. This is why many people on a low-calorie diet make sprouts and baby greens the mainstay of their diet, especially blended for easier digestion.

Many raw food dietitians avoid grains in favor of nuts and seeds. While raw nuts and seeds contain enzyme deactivators that harm the stomach and cooking nuts and seeds denatures their fats and oils, soaked or sprouted nuts and seeds have natural oils and active enzymes at the same time.

One main criticism of the raw foods diet, especially the insistence on the health value of sprouted nuts, seeds, and occasionally grains, is the very intensive labour time require for food preparation. Not only does it take a long time and lots of labour to sprout nuts and seeds, but the sprouting process makes them not as shelf-stable.

Furthermore, unlike most cooked foods, there are very few commercial avenues for purchasing such foods. Most "raw foods bars" are raw but not sprouted (and therefore do not have active enzymes) because the bars would not keep as long on the shelf. The same is true of most nut and seed butters.

A seed is essentially a baby in a suitcase carrying its lunch. "Baby" refers to the embryo, or immature plant, that will grow and develop into the seedling and ultimately the mature plant. The "suitcase" is the seed coat (or testa) that surrounds the seeds and "lunch" refers to the nutritive source for the germinating seedling. The food for the germinating seedling may be stored in part of the embryo itself, such as the fleshy cotyledons of a bean seed, or it may take other forms including endosperm, which is a special starch-rich storage tissue that surrounds the embryo.

A seed is officially considered to have germinated when the young root, called the radicle, emerges from the seed coat. To germinate, a seed requires three things – water, oxygen, and a suitable temperature. Water uptake, also called imbibition, is the first stage of seed germination. During this process the dry seed, which typically has a water content of less than 10%, absorbs water and swells.

This process serves to hydrate the dry components of the seed and active the metabolic machinery necessary for germination. Among the early metabolic activities occurring in the seed is the breakdown of starches stored in the seed into simple sugars that can be used for energy and building blocks for necessary cellular structures.

Except for the first half-hour or so of germination when little oxygen is present, seed germination and subsequent seedling growth requires oxygen. It is required, in large part, for use in the cellular structures, called mitochondria, to produce ATP (energy).

A suitable temperature is necessary to optimize the metabolic reactions required for germination. The seeds of every species have an optimal temperature for germination; some species, such as the gourds and squashes prefer warm temperatures while other species such as radish can tolerate cooler temperatures for germination.

A seed that has not germinated because it is lacking one or more of the necessary requirements for germination is termed quiescent. These seeds are simply "resting", waiting for the appropriate conditions for germination. Given water, oxygen and/or a suitable temperature, a quiescent seed will germinate. However, even if given the proper conditions, a seed may not germinate. These seeds may fail to germinate because the seed is either dormant or "dead".

Dormant seeds have the potential to germinate but are prevented from doing so by some mechanism. Thus, even though all the proper growth conditions are present, they don't germinate unless they have been "primed" and there dormancy mechanism has been overcome. There are many dormancy mechanisms in seeds.

When some seeds, like hemp, are shed they have immature embryos that will not germinate until they undergo a period of development (called after-ripening). Other seeds, like apple, require a cold treatment, called stratification, for germination. Many of our native plant seeds must be stratified. Some seeds have a hard seed coat that needs to be nicked (called scarification) for germination. This usually occurs as the result of natural freeze-thaw cycles. Still other seeds require a period of heat in order to germinate. Many of these species are winter annuals that germinate in the late summer/early fall.

Ultimately, the function of these varied dormancy mechanisms is to enable the seed time to disperse from the parent plant and to avoid germinating during unfavorable weather. Humans have attempted to breed dormancy mechanisms from our crop plants. Although an advantage for a wild plant, dormancy is a problem if a farmer who want the crop to germinate uniformly and immediately upon planting.

It's not easy to tell if a seed is "dead". Only if it fails to germinate when provided the proper conditions and any dormancy mechanisms are broken can we consider a seed "dead".

SEED GERMINATION

Plants produce many fruits and therefore many seeds. This is because only a few of these seeds will find suitable places and conditions in which to germinate. The majority of the seeds will be wasted; some wind-dispersed seeds could fall in water or on stones.

When fruits and seeds are dispersed the seeds have very little water in them. The outer covering, the testa, is hard and sometimes rough and creased. Although the seeds seem dry and inactive they are very much alive. They are said to be in a dormant state and some seeds can remain like this for a long period of time. Inside the testa there is just enough water to keep the embryo (the baby plant) living until the external conditions are correct for the seed to germinate.

The Structure of a Haricot Bean Seed

The outer covering of a seed is called the testa. It is usually hard and protects the softer parts of the seed. There is a tiny hole in the testa called the micropyle. When the seed is ready to germinate, water is taken in through the micropyle. The first root, called the radicle, will grow out of the seed through the micropyle. On the surface of the testa there is a scar which shows the place where the seed was attached to the fruit.

A dried seed is difficult to study. If the seed is soaked in water for twelve hours it takes in some of the water and swells. Peas and beans are useful seeds to study after they have been soaked because they are big. The testa can be removed easily and the inside of the seed can be examined under the binocular microscope.

INSIDE THE SEED

Inside both the pea and the bean there are two large food-storing structures called cotyledons. The food contained in the cotyledons gives energy to the embryo when the seed germinates. The embryo is found between the two cotyledons. It is made up of two parts; the radicle, or first root, which points toward the micropyle, and the plumule, or first shoot. Under the binocular microscope the plumule can be seen as two very tiny leaves.

Some types of seeds have only one small cotyledon. Most cereal grains, such as wheat, are like this. The embryo is surrounded by a structure called the endosperm which is also a food storage area. Both the cotyledon and the endosperm will provide energy to the embryo during germination.

SEED GERMINATION

A seed certainly looks dead. It does not seem to move, to grow, nor do anything. In fact, even with biochemical tests for the metabolic processes we associate with life (respiration, etc.) the rate of these processes is so slow that it would be difficult to determine whether there really was anything alive in a seed.

Indeed if a seed is not allowed to germinate (sprout) within some certain length of time, the embryo inside will die. Each species of seed has a certain length of viability. Some maple species have seeds that need to sprout within two weeks of being dispersed, or they die. Some seeds of Lotus plants are known to be up to 2000 years old and still can be germinated.

Assuming the seed is still viable, the embryo inside the seed coat needs something to get its metabolism actived to start the embryo growing. The process of getting a seed to germinate can be simple or complicated, and this our present subject.

Seeds Lacking True Dormancy

Common vegetable garden seeds generally lack any kind of dormancy. The seeds are ready to sprout. All they need is some moisture to get their biochemistry activated, and temperature warm enough to allow the chemistry of life to proceed. Seeds taken from the wild, however, are frequently endowed with deeper forms of dormancy.

Seeds With Truly Dormant Embryos

There are several mechanisms that permit seeds to be truly dormant.

Thick Seed Coat

Many kinds of seeds have very thick seed coats. These obviously keep water out of the seed, so the embryo cannot get the water needed to activate its metabolism and start growing. The lotus seeds are an example of this. An outstanding example from the northern temperate zone is the Kentucky coffee tree (*Gymnocladus dioica*). The seed coat is perhaps two millimeters thick! You can throw them as hard as you can against a concrete sidewalk and they just bounce! How could such a seed actually sprout?

The Kentucky coffee tree holds its seed pods in the the top of the tree all winter. The inside of the pod is fleshy (lots of water). The pods are very dark in colour.

Other species might use some pounding along a river or drop seeds into seacoast surf to abrade the thick seed coat. Some of the sea beans do this. Other seeds might need an vertebrate or other animal to attack the seed coat (but give up trying to eat the seed) and thereby weaken the coat. The process of nicking the thick seed coat to initiate germination is called scarification.

A final, and very common, example of a way to scarify a seed coat is observed in strawberry and raspberry. The thick seed coat is designed to be swallowed by the frugivore. The animal digests the fruit pulp, but the seed coat passes through the digestive system still protecting the viable embryo inside, but weakened enough to allow sprouting! The seed is deposited with a little organic fertilizer in the environment and can now sprout!

Thin Seed Coat

A thin seed coat is so thin that it is no barrier to water. Some other kind of dormancy mechanism is needed. Knowing that light can penetrate thin layers of plant tissue should give you the idea that light might be a signal. That plants can absorb light and respond biochemically is a fact you know from your study of photosynthesis. All we need is a pigment molecule that can absorb light and cause a change in the behaviour of the embryo.

The pigment is phytochrome. Like chlorophyll, it is made of a chromophore with tetrapyrole structure and is associated with proteins. This pigment is different from chlorophyll, however, in one critical way. It exists in two inter-convertible forms.

One form of phytochrome, named Pfr, is the form of the phytochome found in plant cells that are exposed to red (660 nm) or common white light. This form of phytochrome is biologically very active and plays a role in all systems when a plant needs to know if the lights are "on" or "off." In lettuce (*Lactuca sativa*) seeds, Pfr causes the seeds to begin to germinate as we will soon see. Thus lettuce seeds germinate only when placed in white or red light. Buried in deep soil, they will not germinate.

The other form of phytochrome, named Pr, is formed when phytochrome is exposed to far-red (730 nm) light. This form is biologically inactive or inhibits responses. Thus if lettuce seeds are placed in far-red light they do not germinate.

Large seeds have lots of storage material. If their seed coat is very thin, their evolution may have arrived at a completely different response. Think about pea seeds. They are large and have very thin seed coats. How would they respond to light?

Insufficient Development

If a seed's embryo is not completely developed, some additional maturation may be needed before the seed can sprout. This happens in seeds with little-to-no storage material invested in the seed. Examples include orchid seeds. They are the size of dust and have almost nothing but a very immature embryo on-board.

Such a seed needs an association with fungi in the soil or other environments to feed the developing embryo until the embryo is mature enough to actually penetrate the seed coat. These seeds are also likely to have a very brief viability. The fungal association must be established rapidly or the embryo dies.

Inhibitors Present

Many plant species invest chemicals in the developing seeds, and these chemicals inhibit the development of the embryos. They keep the embryos dormant. Obviously the seed must have some way to eliminate these chemicals before they can sprout.

Abscisic Acid

Many temperate zone species that use inhibitors use abscisic acid. This chemical induces dormancy in the embryo. The chemical is produced in abundance in the late summer and early fall. The seeds in the fruits become dormant so, even if they are dispersed in autumn, they cannot sprout. During the winter enzymes in the seeds degrade the abscisic acid. By spring the abscisic acid is gone and the seed can sprout.

We can collect seeds of these species and get them to sprout early. The seeds are put in moist soil and refrigerated for about four weeks (a process often called stratification). This is sufficient time to degrade the abscisic acid. Then the planted seeds are placed in a warm greenhouse.

The seeds assume winter is over, spring has come, and they begin to sprout. This process is called vernalization. If you think of "vernal" as meaning "spring" then you understand how we got this name!

PHENOLIC COMPOUNDS

Plants that live in deserts have a different problem. There is no cold, moist, winter to allow vernalization of abscisic acid. These plants instead use more potent toxins, phenolic compounds, to keep their seeds dormant until the proper season for germination. Phenolic compounds are freely water-soluble, the plant is living in a desert.

Deserts typically have very long dry seasons and a short wet season accompanied by flash floods and so on. How do you think the phenolic compounds are lost? How would the mechanism ensure that seeds do not sprout in the dry season, but only after the seed could be sure it is in the wet season? The word leaching might give you a hint?

As any "typical" seed it has three fundamental parts:

- A seed coat
- A storage area (in this case the endosperm) and
- A dormant embryo

The seed coat is really a fruit coat. In all grains, the seed coat (former ovule integument) is fused to the ovary wall (the true fruit wall). So in fact, the grain is technically a fruit (caryopsis) even though we often call it a seed.

Barley is used primarily in beer making. Brewers use barley as a source of sugar to make the alcohol for beer. Brewers knew that sprouting the barley seeds improves the sugar yields tremendously, but they wanted even more sugar yield. The brewers came to plant physiologists to find out if anything could be done.

Study of the seed showed that the seed/fruit coat is water-resistant and thereby reduces the rate of water uptake by the seed, and water uptake is essential for seed germination and the improved sugar yields.

The sugar content of a dry barley seed is actually quite low, but the endosperm holds a huge reserve of starch. Starch is a polymer of sugars. In a sense it is a long chain of sugar molecules linked together. This is probably the source of the sugar.

The barley embryo has three parts:

- The cotyledon (or seed leaf). Since there is only one, barley is a monocot.
- The epicotyl (becomes the shoot)
- The radicle (becomes the root)

Seed germination is said to have occurred when growth of the radicle bursts the seed coat and protrudes as a young root. The energy for seed germination probably comes from respiration of the sugar in the endosperm. However, the embryo and the starch are separated from each other. There must be some chemical communications that physiologists could manipulate to enhance sugar yield for the brewers.

The endosperm of the seed has two parts. The bulk of the volume is a starch storage area. The covering layer is called the aleurone layer. The aleurone is made of cells that store protein in abundance.

You probably know about this in the case of rice. Rice grains are treated before we buy them for cooking. The fruit/ seed coat is milled off and discarded as it has little food value. The embryo breaks out and is called the "germ." In wheat, the embryos are sold as wheat germ. The milling process leaves the rice endosperm surrounded by aleurone.

This is sold as brown rice and has food value in starch and protein. It has an interesting texture after cooking. If the endosperm is polished before it is sold, the aleurone can be removed. Polishing results in white rice which has food value in starch only. Thus white rice, while considered much "finer" than brown rice, is actually a lower-quality food.

The first step in barley seed germination is imbibition. In this process, water penetrates the seed coat and begins to soften the hard, dry tissues inside. The water uptake causes the grain to swell up. The seed/fruit coat usually splits open allowing water to enter even faster. The water begins to activate the biochemistry of the dormant embryo.

The water coming into the seed and embryo dissolves a chemical made inside the embryo. This chemical is called Gibberellic Acid (GA). It is a plant hormone, not too different from steroids.

The dissolved GA is transported with the water through the rest of the seed tissues until it arrives at the aleurone layer.

The GA crosses into the cytoplasm of the aleurone cells and turns on certain genes in the nuclear DNA. DNA is, of course, the hereditary molecule and contains the instructions for making every protein needed for the survival of a barley plant.

The precise mechanism of how GA turns on the DNA is unknown at present. It is clear however that the mode of action is to turn on just *certain* genes in the DNA.

The genes that are turned on are transcribed. The information archived in the DNA is precious, so the aleurone cells make a disposable RNA copy of the gene that is turned on. This disposable copy of the information, a kind of blueprint, is often called messenger RNA. The process of making this RNA copy is called transcription.

The RNA that was made in the transcription process is transported into the cytoplasm of the aleurone cells.

In the cytoplasm, the messenger RNA joins up with a ribosome to begin the process of making a protein. This process is often called protein synthesis or translation. In this process, the ribosome examines the information held in the sequence of bases in the RNA.

Transfer RNAs charged with particular amino acids are moved into the position specified by the instructions in the messenger RNA, and the amino acids are joined in a proper sequence by the ribosome. The sequence of amino acids determines the properties of the protein being assembled.

In this case, the critical protein made with the information held in the RNA is amylase. This protein turns out to be an enzyme of great importance.

The process of information held in the genes of DNA being transcribed into RNA and then translated into protein constitutes the central dogma of genetics. You will want to remember later, that it is the signal of Gibberellic acid that initiates this whole chain of events.

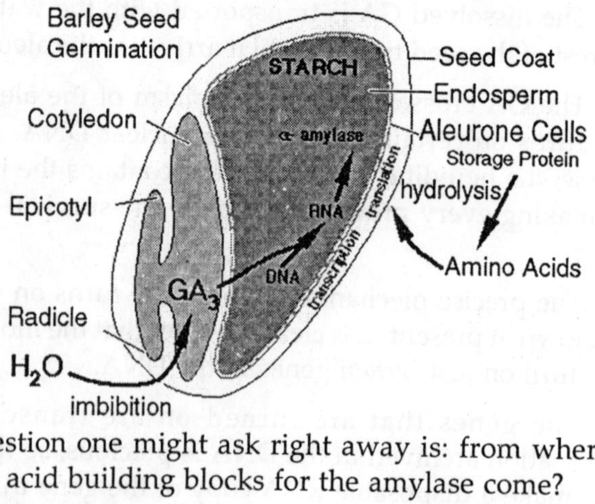

A question one might ask right away is: from where do the amino acid building blocks for the amylase come?

The answer is: from some other biochemistry in the aleurone cells. This biochemistry causes the storage proteins in the aleurone cells to be digested by hydrolytic enzymes. The hydrolysis is accelerated by enzymes known as proteases. These enzymes increase the rate at which the storage protein is cut into individual amino acids.

The amino acids released by the hydrolysis are then free to be reassembled by the ribosomes into the structure of amylase.

The same thing happens in people. You are not what you eat! You do not slowly become a steer by eating beef. The protein in your hamburger is digested into amino acids. Those amino acids are then reassembled into human proteins. Since the instructions for reassembling the amino acids come from your human DNA, the proteins produced are human, not bovine!

In a similar way, barley aleurone storage proteins are digested and amylase is made from the released amino acids.

Amylase is not just any old protein. It happens to be an enzyme. It catalyzes (accelerates) a particular chemical reaction. Specifically, the amylase speeds the hydrolysis of starch into its component sugar units.

Hurrah! The physiologists now know how the sugar is made in a seed. They can also tell the brewers how to increase the sugar yields for beer making.

Note that Gibberellic Acid was the first plant hormone purified in large quantities to be successfully used to increase the yield of any economically-important biochemical in plants!

Remember that we wanted to get the seed germinated (sprouted). This does not occur until after the released sugar is transported from the endosperm to the embryo. The cotyledon is a major sugar transfer area in grains. Active sugar uptake is its specialized function. The sugar is transported into the embryo and used as a fuel and building block for growth of the embryo itself.

The resulting embryo growth will include the emergence of the radicle from the seed/fruit coat. Germination is thus complete! Almost the end of the story...

Our barley seed story is great, but it does not explain some other forms of seed germination.

The next great finding in the study of seed germination is what happens in lettuce seeds (*Lactuca sativa*). Lettuce seeds look somewhat different from barley seeds, because they are dicots. They also have separate seed coat and fruit walls. But the things you buy as lettuce "seeds" are really lettuce fruits (called cypselas). As different as they are, the seed germination process in lettuce needs just one step more than barley to get its seeds germinating.

Lettuce cypselas have very thin fruit walls and seed coats, so light can penetrate these coverings. As is typical of small seeds, light is needed to stimulate seed germination. One might wonder how this would fit into what we just found out with barley.

The DNA must first be photoactivated before GA can have its effect. This is accomplished by the formation of a chemical signal known as Pfr.

Pfr is one of two possible chemical forms of a molecule known as phytochrome. Phytochrome is a pigment that absorbs light energy. When it does absorb light energy its internal chemical structure is altered instantly. Thus Pfr can be converted to the alternate Pr form of phytochrome by simply hitting the seeds with a flash of far-red light (730 nm). Similarly, the Pr form of phytochrome can be converted back into Pfr with a flash of red light (660 nm).

Dry lettuce seeds have slightly more than half of their phytochrome in the Pfr form. So if they are moistened and kept in the dark, some of the seeds will actually germinate.

If we shine red (or it turns out even white) light on imbibing seeds, then virtually all of the phytochrome is converted to the Pfr form and this photoactivates the genes in the DNA. Almost all of the seeds will germinate in this light!

On the other hand, if we shine far-red light on imbibing seeds, then virtually all of the phytochrome in the seeds is converted to the Pr form and none of the seeds will germinate.

Thus phytochrome and light tell lettuce seeds when they are exposed to sunlight and can begin to germinate. Small seeds, such as those in lettuce, do not have reserves enough to start germinating without being able to quickly get to light. It is no surprise that small-seeded species have evolved this mechanism for photoactivation of seed germination. It prevents a deeply-buried seed from trying to germinate!

Chapter 3

Seed Germination

Seed germination is the process of growth of the embryo into a functional plant. It involves the reactivation of the metabolic pathways that lead to growth and the emergence of the radicle or seed root and plumule or shoot.

Three fundamental conditions must exist before germination can occur.

1. The embryo must be alive, called seed viability.
2. Any dormancy requirements that prevent germination must be over come.
3. The proper environmental conditions must exist for germination.

Seed viability determines the percentage of possible seed germination and is affected by a number of different conditions. Some plants do not produce seeds that have functional complete embryos or the seed may have no embryo at all, often called empty seeds. Predators and pathogens can damage or kill the seed while it is still in the fruit or after it is dispersed.

Environmental conditions like flooding or heat can kill the seed before or during germination. The age of the seed affects its health and germination ability, since the seed has a living embryo, over time cells die and cannot be replaced. Some seeds can live for a long time before germination, while others can only survive for a short period after dispersal before they die.

Seed vigor is a measure of the quality of seed, and involves the viability of the seed, the germination percentage, germination rate and the strength of the seedlings produced.

The germination percentage is simply the proportion of seeds that germinate from all seeds subject to the right conditions for growth. The germination rate is the length of time it takes for the seeds to germinate. Germination percentages and rates are affected by seed viability, dormancy and environmental effects that impact on the seed and seedling. In agriculture and horticulture quality seeds have high viability, measured by germination percentage plus the rate of germination.

This is given as a percent of germination over a certain amount of time, 90% germination in 20 days. Many plants produce seeds with varying degrees of dormancy, and different seeds from the same fruit can have different degrees of dormancy. It's possible to have seeds with no dormancy if they are dispersed right away and do not dry (if the seeds dry they go into physiological dormancy). There is great variation amongst plants and a dormant seed is still a viable seed even though the germination rate might be very low.

Environmental conditions effecting seed germination include; water, oxygen, temperature and light.

Three distinct phases of seed germination occur:

1. Water imbibition;
2. Lag phase;
3. Radicle emergence.

In order for the seed coat to split, the embryo must imbibe (soak up water), which causes it to swell, splitting the seed coat. However, the nature of the seed coat determines how rapidly water can penetrate and subsequently initiate germination. The rate of imbibition is dependent on the permeability of the seed coat, amount of water in the environment and the area of contact the seed has to the source of water.

For some seeds, imbibing to much water to quickly can kill the seed. For some seeds, once water is imbibed the germination process can not be stopped and if the seed dries out again it is fatal. While other species have seeds that can imbibe and lose water a few times with out causing ill effects to the seed or drying can cause secondary dormancy.

Inducing Germination

A number of different strategies are used by gardeners and horticulturists to break seed dormancy.

Scarification

Scarification of hard seed coats involving the breaking, scratching or softening by chemicals like acids. Other means of scarification include soaking in hot water or poking holes in the seed with a pin. Sometimes fruits are harvested while the seeds are still immature and the seed coat is not fully developed and sown right away. Under natural conditions the seed coats can be broken by rodents chewing on the seeds, rubbing against rocks or freezing and thawing of surface water, battering on rocks in a stream-bed, or passing through an animal's digestive tract.

In the latter case, the seed coat protects the seed from digestion, while perhaps weakening the seed coat such that the embryo is ready to sprout when it gets deposited (along with a bit of fertilizer) far from the parent plant. Microorganisms are often effective in breaking down hard seed coats and are sometimes used by people as a treatment, the seeds are stored in a moist warm sandy medium for several months under non-sterile conditions.

Stratification

Stratification also called moist-chilling is a method to break down physiological dormancy and involves the addition of moisture to the seeds so they imbibe water and then the seeds are subject to a period of moist chilling to after-ripen the embryo.

Sowing outside in late summer and fall and allowing to overwinter outside under cool conditions is an effective way to stratify seeds, some seeds respond more favorably to periods of osculating temperatures which are part of the natural environment.

Leaching

Leaching or the soaking in water removes chemical inhibitors in some seeds that prevent germination. Rain and melting snow naturally accomplish this task. For seeds that are going to be planted for gardens, the use of running water is best but frequent changes of water are effective too.

Normally 12 to 24 hours of soaking is sufficient, longer soaking especially in stagnant water that is not changed can result in oxygen starvation and seed death. Seeds with hard seed coats can be soaked in hot water to break open the impermeable cell layers that prevent water intake.

Other methods used to assist in the germination of seeds that have dormancy include prechilling, predrying, daily alternation of temperature, light exposure, potassium nitrate, the use of plant growth regulators like gibberellins, cytokinins, ethylene, thiourea, sodium hypochlorite plus others.

ORIGIN AND EVOLUTION LEACHING

The origin of seed plants is a problem that still remains unsolved. However, more and more data tends to place this origin in the middle Devonian. The description in 2004 of the proto-seed *Runcaria heinzelinii* in the Givetian of Belgium is an indication of that ancient origin of seed-plants. As with modern ferns, most land plants before this time reproduced by sending spores into the air, that would land and become whole new plants.

The first "true" seeds are described from the upper Devonian, which is probably the theater of their true first evolutionary radiation. The seed plants progressively became one of the major elements of nearly all ecosystems.

ECONOMIC IMPORTANCE

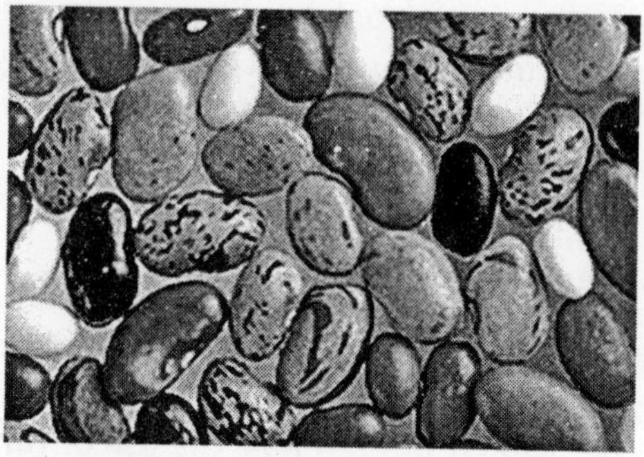

Fig. A variety of bean seeds.

Edible Seeds

Many seeds are edible and the majority of human calories comes from seeds, especially from cereals, legumes and nuts. Seeds also provide most cooking oils, many beverages and spices and some important food additives. In different seeds the seed embryo or the endosperm dominates and provides most of the nutrients.

The storage proteins of the embryo and endosperm differ in their amino acid content and physical properties. The gluten of wheat, important in providing the elastic property to bread dough is strictly an endosperm protein.

Seeds are used to propagate many crops such as cereals, legumes, forest trees, turfgrasses and pasture grasses.

Seeds are also eaten by animals, and are fed to livestock. Many seeds are used as birdseed.

Poison and Food Safety

While some seeds are considered by some as healthy to eat, other seeds may be harmful or poisonous, Plants and seeds often contain chemical compounds to discourage herbivores and seed predators. In some cases, these compounds simply taste bad (such as in mustard), but other compounds are toxic, or breakdown into toxic compounds within the digestive system.

Children, being smaller than adults, are more susceptible to poisoning or death by plants and seeds. One should be satisfied with reliable food safety information before choosing to eat any particular seeds.

An infamously deadly poison, ricin, comes from seeds of the castor bean. Reported lethal doses are anywhere from two to eight seeds, though only a few deaths have been reported when castor beans have been ingested by animals.

In addition, seeds containing amygdalin; apple, apricot, bitter almond, peach, plum, cherry, quince, and others, when consumed in significant amounts, may result in cyanide toxicity. Other seeds than contain poisons include annona, cotton, custard apple, datura, uncooked durian, golden chain, horse-chestnut, larkspur, locoweed, lychee, nectarine, rambutan, rosary pea, sour sop, sugar apple, wisteria, and yew. Another seed poison is strychnine.

The seeds of many legumes, including the common bean (*Phaseolus vulgaris*) contain proteins called lectins which can cause gastric distress if the beans are eaten without cooking. The common bean and many others, including the soybean, also contain trypsin inhibitors which interfere with the action of the digestive enzyme trypsin. Normal cooking processes degrade lectins and trypsin inhibitors to harmless forms.

HOW TO GERMINATE SEED

GENERAL RULES OF SEED SOWING

Seeds need water and oxygen to germinate, so are best started in a light, loose soil that will not compact, get soggy, or crust over. Free flow of water & air are a must. Cover seed with 2 - 4 times their thickness of soil, unless they require light to germinate. Sow shallowly in cold wet spring, more deeply in warm dry summer. Large seed can be soaked overnight and planted singly.

Barely cover small seed, and sprinkle fine seed on the surface and water by misting. Plant flat seed edgewise and winged seed with wing uppermost or broken off. Sowing too thickly wastes seed and weakens the crowded seedlings, but some kinds sprout best if crowded. Lightly tamp soil to insure good contact with the seed, unless heavy. Keep soil moist, not soggy, and do not allow to dry out.

Common causes of failure are soil too heavy, wet or cold, or allowed to dry out, not giving slow seeds long enough to come up, pests eating the seeds or seedlings, and not giving dormant seeds the proper pretreatment. Careful attention to the instructions in the catalog and on the packet will help insure good results.

Common causes of seedling loss are damping off due to poor air circulation & overwatering, drying out or burning due to placing in full sun or outdoor conditions too quickly, transplanting shock (best done on a cool, moist day), and predation by insects, slugs and snails at night.

TEMPERATURE

Most seeds germinate best at warm (70°F) temperatures. Plants from temperate regions, the arctic, high mountains and high deserts often germinate best at cool temperatures. Plants from winter-rain areas like California, the Mediterranean, Chile, S. Africa and parts of Australia also like cool temperatures.

Warm temperatures will often speed germination of these seeds, but lower vigor & survival. Warm desert plants and tropicals like warmth. Temperatures used in the catalog are: Cold (34 - 45°F), Cool (50 - 65°F), Warm (65 - 80°F) and Very Warm (80 - 100°F). When no temperature is given, warm (70°F average) is indicated.

TIME UNTIL GERMINATION

Average range of time to germinate is usually given in weeks. A seed that takes 2 - 3 weeks will usually come up fairly evenly; one that takes 1 - 12 weeks will tend to straggle in irregularly.

Time varies with temperature, so expect considerable variation. Don't give up too soon — many who have given up and sown another seed in the pot end up with two types of plants in the same pot!

Hardy Annuals (HA)

These will stand some frost and can be sown direct to the garden as early as the ground can be worked, March to June. Prepare the soil until a smooth, fine surface is obtained. An attractive annual border can be had by planting in large, irregular drifts.

Half Hardy Annuals (HHA)

These are killed by frost and should be sown in late spring after danger of frost. For early bloom, start early indoors & plant out after danger of frost.

Tender Annuals (TA)

These need warmth and shelter and are best started in pots or flats and planted out in favored spots after the soil has warmed.

Biennials and Winter Annuals

Biennials are sown like half hardy annuals or perennials in spring or fall and planted out in September and October. Winter annuals such as some Californian and desert plants may be grown in summer, but are at their best sown in fall, even if grown in the greenhouse in cold winter areas.

Hardy Perennials (HP)

Many of these germinate readily at warm temperatures, and can be sown direct to the garden or early in the greenhouse or cold frame. If started early, they often bloom the first year. Others germinate best at cool or cold temperatures and the seedlings need cool temperatures. Many have various dormancies & need the pretreatment indicated.

Hardy Trees and Shrubs

Many will germinate readily if sown in spring. Others need cold or other pretreatment, and some are best sown in fall or winter and covered with a mulch or snow. The addition of some forest soil or litter from below both hardwoods and conifers is often very beneficial, greatly increasing seedling growth due to beneficial mycorrhizal inoculation.

Tropical and Subtropical Plants

These often need very warm temperatures. Bottom heat and constant moisture are often beneficial. Some take surprisingly long times to germinate. Many people think all tropical seeds are short-lived and perishable, but many of the longest-lived seeds are tropical, and some even need to be aged dry for a year before they will germinate!

Aquatic Plants

We have had the best results with some of these by sowing in pots, pans, or petri dishes, and submerging so that the soil is under 1/4" of water.

Fern Spores

Spores should be scattered thinly on the surface of a light, sterile, peaty mix which has been watered with clean sterile water about 4 hours before sowing. Cover pot with a pane of glass or enclose in a plastic bag and place out of direct sunlight. A light green film will form on the surface in 1 - 4 weeks or so, up to 6 months for slow species.

Botrychiums may take up to 8 years, and one *Equisetum* takes 18 years! Small round or heart-shaped prothallia will then form. Begin to mist lightly with distilled water weekly to enable fertilization of the egg cells, until true fronds develop. Transplant sporelings when large enough to handle. Good ferns can usually be had in 8 - 12 months from spores. Ferns are hybridized by sowing 2 kinds of spores together.

Soil Mixes

Many seeds do well sown direct to ordinary garden soil, but even good soil may be poor in pots or flats. These need a lighter, looser soil. Most commercial mixes are fine, but the addition of some garden soil and compost will often insure adequate beneficial micro-organisms & fungi.

A good soil mix can be made at home from 1/3 garden loam, 1/3 peat or compost, and 1/3 gritty sand. Number 1 limestone chick-grit makes a good top dressing for many alpines or slow germinating seed to discourage algae growth. Crushed charcoal also helps.

Hard Shelled Seed

These have hard impermeable shells and need nicking or scratching the coat to allow water to enter and the seed to germinate. The best results are from the least amount of nicking that will allow water to enter and the seed to swell. Many failures are due to over-nicking and damaging the seed.

Different seeds need varying amounts of nicking. Most do best with lightly rubbing on sandpaper or a file until just the very outer coat is scratched. Often just scratching with a knife-point or scriber works. Others need serious nicking, sometimes with a hacksaw until the white interior shows.

Nicking seed one by one can be tedious but is most effective. Larger lots can be rubbed between two boards covered with sandpaper, or shaken or tumbled in a can lined with sandpaper. Then soak the seed overnight and any that don't swell or soften should be re-nicked and soaked again until swollen.

Nick & Soak means nick first, then soak, and re-nick any hard (unswollen) ones.

Soak & Nick Unswollen means the seed is usually only partly hard and many will swell without nicking. Soak first, nick only the hard ones.

Other Methods for large lots include:

Hot Water Soak: The seeds are placed in a cup and not quite boiling (200°F) water is poured on them and allowed to cool & the seeds to swell. At least 4 - 5 times their volume of water is used. Or the seeds are dipped in boiling water for 10 seconds to 3 minutes. Produces erratic results, often killing many of the seeds and increasing fungal attack.

Dry Heat: The seeds are baked dry in an oven at 140° to 220°F for 4 - 10 hours, or are microwaved for 30 seconds to 4 minutes. Gives erratic results.

Sulfuric acid or lye soaks work with some seeds.

Cold Treatment (Prechill)

Many seeds need a cold moist period before they will sprout. The essentials are moisture, air, cold and time. Dry cold (a packet in the refrigerator) won't work. Often the seed may be sown in fall or winter and allow natural cold & snow to work.

1. Soak seed overnight until swollen or soft (up to four days for large hard nuts). Nick if needed.
2. Mix seed with about 3 times its volume of damp peat moss or vermiculite and place in a plastic bag or pill bottle. Small amounts may be conveniently layered between damp paper towels or coffee filters. Remember, air is essential; avoid sogginess. Label the bag with the name of the seed and date to be removed from cold. Mark your calendar too!
3. Store in the refrigerator (34 - 40°F) for the time specified in the catalog. Four to 12 weeks is usual.
4. Remove the seed and sow. Seed is best kept cool (50°F or so) for a week after sowing, and gradually brought up to warm temperatures. Warming too quickly can be fatal for some seeds.

'Cold germinators' only sprout at cold temperatures, and are most conveniently treated by layering in damp towels and pricked off as they germinate.

If using outdoor treatment, hold pots at above freezing for a few weeks before putting outside.

Snow cover is often highly beneficial.

Dormancy is highly variable. Sometimes a seed collected in the warm lowlands will germinate readily, but the same species collected at a high elevation will need cold. Dormancy even varies between individual plants at the same site, and varies with weather before harvest and conditions of storage after harvest.

Gibberellic Acid-3 (GA-3)

GA-3 is a naturally occurring plant growth regulator. Presoaking seeds in GA-3 will often cause rapid germination of many highly dormant seeds that would otherwise require cold, aging, light or other prolonged treatments.

Many different types of dormancy are overcome. You may buy our 'GA-3 Primed' seed which are ready to plant, or the GA-3 powder or one of our kits. GA-3 is safe, easy to use, economical and rapidly becoming a standard tool for germinating seeds.

Other Treatments

Dry Storage or Ageing: Many seeds will not germinate when freshly harvested, but are dormant until after a period of dry storage ranging from 1 - 12 months or up to 5 years. The time varies with temperature, humidity and oxygen. We try to supply aged seed whenever necessary. Often this type of dormancy can be broken with GA-3.

Warm Moist Treatment: Many seeds need 1 - 4 months of warm moist treatment, followed by cold treatment to sprout. In some, the root sprouts during the warm period, but the shoot does not sprout until after a cold period (epicotyl dormancy). Called 'two-step' germinators by Deno.

Some of these have unformed embryos which must first develop at warm temperatures. Some seeds are '70 - 40' germinators, which need warm moist, then cold moist, and germinate at cold temperatures. Some tropicals which are slow to germinate (many palms and aralias) need prolonged warmth to first develop their embryos. Many seeds are 'multicycle' germinators, needing several cycles of cold and warm periods.

Soaking, Washing or Leaching: Many seeds contain germination inhibitors which in nature prevent the seed from germinating except during the wet season (tropicals) or only after sufficient rain has fallen (desert plants). Soaking the seed in shallow water and changing every day for several days will leach out inhibitors. Heavy daily watering of pots may work. Some have oily coats and need detergent or peroxide soaks.

Light or Darkness: Some seeds need light to germinate; others need darkness, and light prevents sprouting. If light is required, sow on the surface; if darkness is needed, cover seed well.

Oscillating Temperatures: Some seeds need the daily change from warm to cold that occurs in spring and fall, and they germinate best sown in pots and exposed to outdoor conditions.

Other Treatments: Soaking seed in potassium nitrate (KNO3), hydrogen peroxide, citric acid, sodium hypochlorite (bleach), smoke solution, charred wood leachate, or enzyme solutions have all been used to trigger germination of seeds.

Smoke Treatment

Smoke treatment often helps germination of plants from fire-prone environments, particularly Mediterranean-climate plants such as many species from California, Chile, Australia, South Africa, and the Mediterranean region.

Prepare a dilute smoke solution by adding one part commercial smoke flavouring to nine parts water. Either soak the seeds in this solution overnight (or until they swell), or water the pot or flat once with this solution. Smoke flavouring is found in the spice and flavouring or barbecue section of the grocery store. It comes in small brown bottles of liquid, called "liquid smoke" or "hickory seasoning". Look for the "all natural" type that lists only water and natural smoke concentrate as ingredients. "Wright's" is the brand we find out here in California.

Smoke treatment is still experimental, and you may have to try different dilutions. Let us know your results! Another method is to burn a layer of straw or pine needles on top of the flat.

PLASTIC BAG METHOD

Best for very slow to germinate seeds, very tiny dust-like seeds that can't be allowed to dry out, and very slow-growing seedlings. Small clean pots are filled with damp, sterile, soilless mix. Milled sphagnum or a light dusting with powdered charcoal discourage fungi & algae. Seeds are sown and the whole pot is sealed in a plastic bag and placed out of direct sunlight.

This creates a mini greenhouse and the soil will not dry out and the seeds are protected from mice, etc. Pots can be left for years with no care other than regular checking for seedlings. As soon as seedlings appear, begin hardening off. Bagged pots may be kept under fluorescent lights without overheating. Don't forget to label!

Hardiness Zones

The rating of plants according to the hardiness zones in which they will grow has improved greatly over the last twenty years. They are fine for common, widely grown plants, but for most plants, our knowledge of their ecological limits is incomplete. Too often, the assignment of zones is merely educated guesswork.

Many plants will do well far outside their usual ecological range, or seeds collected at the species northernmost limit will be much hardier than usual. Also, a heavy mulch, or planting by a south-facing wall, or other simple tricks can greatly extend a species range in cultivation. So don't rely too much on the zone numbers, or let them prevent you from experimenting!

The Limits of the average Annual Minimum Temperatures for each Zone

Zone 1: Below -50°F

Zone 2: -50° to -40°F

Zone 3: -40° to -30°F

Zone 4: -30° to -20°F

Zone 5: -20° to -10°F

Zone 6: -10° to 0°F

Zone 7: 0° to 10°F

Zone 8: 10° to 20°F

Zone 9: 20° to 30°F

Zone 10: 30° to 40°F

SEED LONGEVITY

Many people have the mistaken idea that seed must be very fresh to grow, and the occasional failure of a packet is blamed on 'old seed'. Properly stored, seed of most species is viable many more years than most people would expect, some giving high germination for decades. Though some species are short-lived, many others are dormant when freshly harvested, and actually need an aging period of weeks, months, or even years before reaching full germination.

Therefore the longevity of various seeds are included. And all of seeds are stored in glass jars in a low humidity atmosphere, ideal for preserving viability. Should you find you need to store some seed till next season, keep the packets in a well closed glass jar in a *cool dry place.*

Seed *half-life* is frequently given in the catalog. This is the average number of years it takes for seed of that species to drop to half of maximum germination, when stored under ordinary conditions.

"In several cases, maximum germinability was not attained until after some years of storage." —D. A. Priestly, in describing a longevity study of 36 species of common seeds. It was found that 14 of the 36 species retained half their viability for *ten years or more of open storage.* —J.L.H., 11/87.

About Difficult Seeds

Over millions of years, wild plants have evolved germination strategies which ensure their survival, but which may not be convenient for the home gardener who wants a quick and even stand of plants from a packet of seed. Many seeds sprout irregularly, so that if the first flush of seedlings is killed by adverse weather, insect predation, etc., more will come along to take their place. In adaptation to various environments, some seeds need periods of cold, warmth, darkness or light, fire, etc.

Some have seedcoats of varying hardness or impermeability, and others contain chemical germination inhibitors which must be leached from the seed before it can sprout. Some species disperse themselves over wide areas by being eaten by animals, the seed sprouting far from the mother plant, the seedcoat softened by digestive juices. Many seeds have internal clocks, and give much higher germination at certain times of the year, regardless of the treatment given.

All seeds wait for the correct time and conditions before sprouting, and the gardener must mimic those conditions to ensure successful germination.

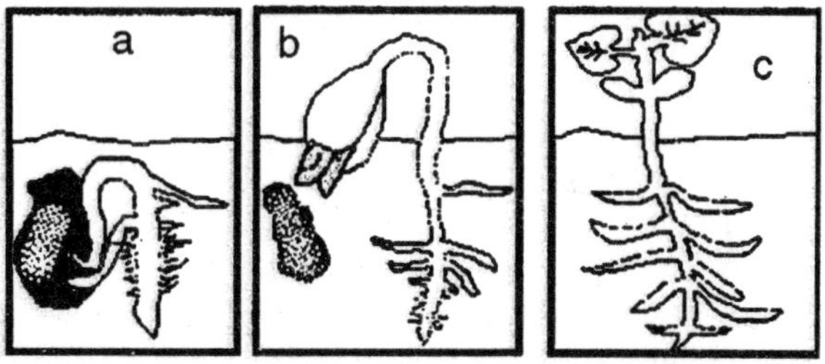

Fig. (a) First the young root appears. (b) The skin splits and the shoot straightens, pulling the cotyledons with it. (c) The main root gets bigger and grows side roots, and the first leaves appear

We are continually testing seed for germination, and conduct research into improving methods of handling difficult species. For slow seeds, which take months or years, making a standard test impractical, we may use 2,3,5-triphenyltetrazolium chloride to test for hydrogenase enzyme activity, which quickly indicates whether there is a living embryo in the seed. It can take years of research to determine the best ways to germinate a specific species.

Oryzopsis seed has been studied for over 50 years, yet we still do not fully understand its requirements. Even then, seed collected from one population may germinate readily, yet the same species gathered in a cold-winter area may need cold treatment.

We offer many seeds which are easy, and sprout quickly and evenly. But with some you must be prepared to experiment, be patient, and use your initiative and intuition. Remember that with some rare species, you are venturing into unknown territory. Most corporate seed companies will not carry difficult seeds, only selling, easy, mass-produced varieties.

Chapter 4

Seed and Fruit Development

Senescence - An introduction to senescence involves the terms monocarpic and polycarpic behaviour. Some plants die at the time when their fruits mature i.e. cereals, peas, soybeans, etc. and this is called monocarpic senescence. This behaviour is distinguished from polycarpic species which go through many cycles of growth and reproduction before death. Cereals are characterized by having all vegetative apices transformed into determinate reproductive structures.

Others like soybean show a close correlation between seed-filling and senescence and this lead to the Nutrient Diversion Hypothesis. This hypothesis is based largely on the observation that removal of flowers or developing fruits will delay senescence. There are many exceptions to this hypothesis and it is an over-simplification, but useful in thinking about mechanisms. ABA applied to leaves of fruiting soybean plants accelerated their senescence, but did not enhance the senescence of depodded plants.

Thus ABA does not seem to be the senescence-inducing factor. It is in fact carried preferentially from leaves to fruits. The role of hormones in regulating senescence usually involves two aspects:

1. Elimination of sources of auxin as vegetative shoot apices are converted to flower development and young leaves mature;
2. A hypothetical factor for senescence (SF) which is produced by developing fruits and promotes senescence.

In addition, cytokinins perpetuate the vegetative condition and also decline with auxin. Developing fruits not only produce SF, but also ethylene.

ETHYLENE

The ethylene biosynthetic pathway that the two important enzymes are ACC synthase and ACC oxidase. Also that ACC synthase is encoded by a multi-gene family, and certain members are expressed during fruit ripening. Ethylene is involved not only in senescence, but also in germination, cell elongation, fruit ripening, and other aspects of growth and development.

In Arabidopsis ethylene sensing involves a protein kinase transduction pathway. This pathway has two genes, one encodes a protein with similarity to the Raf family of Ser/Thr protein kinases, the other most likely encodes an ethylene receptor that functions in a pathway similar to the prokaryotic two-component histidine kinase system. The triple response of seedlings grown in the dark is an ethylene response, and mutants that fail to display this response, even in the presence of exogenous ethylene, are likely to be unable to produce the ethylene receptor.

These ethylene insensitive mutants appear as tall seedling above a lawn of short wild-type ones. The ETR1 gene encodes a membrane-associated disulfide-linked dimer protein that is suggested as the receptor for ethylene. Review the other mutants which are defective in the regulation of ethylene biosynthesis, or at various places in the transduction pathways leading to the many ethylene responses. A number of auxin resistant mutants also display resistance to ethylene.

SEED DEVELOPMENT

A generalized representation of embryo development in angiosperms from zygote to seed usually includes: embryogenesis; maturation; and desiccation. Embryo development is accompanied by reserve food accumulation, and the synthesis of seed storage proteins is the most intensively studied. The maturation loop can be bypassed in culture by precocious germination of embryos.

This occurs in the absence of ABA and the presence of reduced nitrogen and carbon (i.e. glutamine & sucrose). In the presence of high osmoticum and/or ABA, maturation is promoted and precocious germination inhibited. Upon desiccation, which often involves a water content change of 85% to 10%, these processes are finalized, and normal germination will occur in response to imbibition.

In culture, embryos can reversible enter or leave the maturation loop by application or removal of high osmoticum or ABA. The role of various hormones during embryogenesis has been studied and can be summarized as follows: early stages are promoted by low concentrations of auxin, inhibited by high concentrations, CKs enhance the cell cycle especially in endosperm tissue, both GAs and CKs play a role in suspensor development and supply hormones to the embryo.

Mid-stages show high levels of both IAA and GAs during cell expansion and fruit growth, mutants suggest that low levels of GAs are more effective, ethylene has a role in seed germination rather than during seed development; late embryogenesis shows high levels of ABA and decreasing levels of GA and IAA as desiccation occurs, ABA decreases in dry seed and inhibits germination, ABA-insensitive mutants exhibit vivipary and fail to develop ABA-inducible proteins, the role of ABA in development is not clear and it is thought that it plays no role in dicot maturation programme.

Seed Storage Proteins - seed storage proteins are synthesized in large amounts and occur in protein bodies in cotyledon or endosperm cells. Those found in aleurone cells are different and contribute little to storage proteins. Each plant produces a characteristic array of storage proteins, which are classified on such characteristics as size, solubility, and amino acid composition. They are encoded by small gene families and mutations have major effects on their amounts and amino acid composition.

The mutation opaque-2 in maize seems to be regulatory and controls a cluster of structural genes. The gene encoded phaseolin, a glycoprotein constituting up to 50% of the storage proteins in bean. Since this paper a number of seed storage protein genes have been cloned and used to make transgenic plants.

SEED GERMINATION

review the general aspects of seed germination including the imbibition of water, cell expansion, hydrolysis and transport of food reserves in endosperm or cotyledons. The food reserves may include starch, fat, protein, or some combination, plus other compounds like phytin, which contains phytic acid, a hexaphosphate of inositol where the ring hydroxyls have become esters of phosphoric acid, then K^+, Ca^{2+}, or Mg^{2+} salts.

There are many different types of dormancy including immature embryos and those requiring an after-ripening period, various characteristics of the seed coat like hard and impermeable, a variety of light effects, and a variety of temperature requirements. Review the viviparous mutants of Arabidopsis and their role in dormancy and the control of germination. Seeds that store lipids usually require the biogenesis of the glyoxysome for gluconeogenesis.

The famous Rapp & Randall scheme for the subcellular and metabolic reactions of fatty acid breakdown and the synthesis of sucrose and new fatty acids in castor bean endosperm. The complex interplay between glyoxysomes, mitochondria, proplastids, and the cytoplasm, as well as key enzymes like isocitric lyase (de novo in many species), malate synthase, and others, make this an interesting and complex story. This would also be a good time to review the GA/alpha amylase etc. story in barley aleurone cells.

FRUIT RIPENING

Fruit ripening involves the events that lead to changes in the colour, softness, aroma, and flavour in various fruits. Some common features of these processes include the biosynthesis of ethylene and especially the concept of two systems for ethylene production. Fruits can be classified according to whether they are climacteric or nonclimacteric.

System I ethylene biosynthesis is common to climacteric and non-climacteric fruits, and System II ethylene biosynthesis is found only in climacteric fruits and is termed autocatalytic since after exposure to ethylene there is a stimulation of further ethylene production. This induction results in massive increases in ethylene production. Ethylene is the central factor in fruit ripening, ripening fruit produce ethylene and are in turn induced to ripen by ethylene.

Thus to delay or inhibit ripening a block in ethylene production in vivo should work. Various antisense genes that inhibit synthesis of ethylene were studied and transformed tomato plants with a antisense ACC oxidase gene synthesized far less ethylene that wild type and fruit ripened very slowly. The reading list, which gives the details of how transformed tomato plants with an antisense ACC synthase gene specific to an ACC synthase enzyme that is expressed in ripening fruit was shown to suppress ethylene synthesis, some 99.5%.

Such transgenic fruit remained unripe until treated with exogenous ethylene, when they ripened into normal fruit. Thus ethylene appears to be the inducer for ripening and not just some by-product of ripening. Other constructs that had various controls on ethylene biosynthesis also were shown to work, including an ACC deaminase from soil bacteria.

These transgenic plants also allowed the analysis of other genes and their products for their role in ripening. One example is polygalacturonase, a key enzyme in fruit softening, which had normal expression in the antisense ACC synthase transgenic plant, and thus may not be solely responsible for the softening response in ripening.

Fruits develop from ovaries and this development is triggered by pollination or fertilization. In fleshy fruits, such as tomato, the bulk of the fruit is derived from L3. Only the epidermal and subepidermal layers are derived from L1 and

L2. A good review of this process is found in Gillaspy, J. et al. 1993. Fruits: a developmental perspective.

The development and ripening of commercial fruit have been staged for growers, processors, and sellers, and has been divided into three phases. Phase I occurs at anthesis and involves ovary development, fertilization, and fruit set. Cell Cycles occur in Phase II when seeds form and embryogenesis occurs. In Phase III, cells of the fruit expand and the embryos mature. The onset of cell expansion of tomato fruit growth coincides with a peak in auxin accumulation.

ABA - This seems like a good time to review ABA signal transduction in seeds, since this was the subject of a recent review in the ARPPPMB. Endogenous ABA content peaks during the last 2/3 of seed development, and then lowers in the dry seed. ABA has been shown to regulate the accumulation of food reserves, the acquisition of desiccation tolerance, and the induction of seed dormancy. It is known that exogenous ABA inhibits precocious germination of immature embryos in culture, and that in some cases of vivipary the seeds are not able to synthesize normal amounts of ABA, and that both cases are related to ABA biosynthetic mutants.

Thus it is clear that endogenous ABA inhibits precocious germination and promotes seed dormancy. Genetic analysis of signaling elements that mediate regulation of seed dormancy and germination has revealed a variety of mutations that alter sensitivity to ABA in both maize and Arabidopsis. Several of these genes have been cloned and studied. Some vivipary embryos do not have reduced ABA content, but rather a reduced sensitivity to germination inhibition by exogenous ABA in culture.

One such gene is VP1 of maize which encodes a seed-specific transcription factor. The ABA-insensitive mutants of Arabidopsis ABI1, ABI2, etc. were identified by selecting seeds able to germinate in the presence of 3-10 micromolar ABA, a concentration which is inhibitory to wild type seeds.

These mutants also exhibit a marked reduction in seed dormancy.

Other mutants termed ERA or Enhanced Response to ABA types were identified by lack of germination in the presence of low concentrations of ABA (0.3 micromolar) that are not inhibitory to wild type germination.

The accumulation of reserve foods and the acquisition of desiccation tolerance have also been subjected to mutational analysis. The expression of specific sets of mRNA encoding storage proteins and late-embryogenesis-abundant (LEA) proteins encoding factors involved in the acquisition of desiccation tolerance can be induced by exogenous ABA in cultured embryos. ABA-deficient mutants also support similar activities.

Two classes of transcriptional elements have been identified: one class is Cis-acting promoter elements involved in the activation of gene expression by ABA; another class includes Trans-acting factors and upstream signaling elements. Thus there is beginning an accumulation of genetic evidence that may allow the elucidation of the transduction pathway(s) for ABA in the control of the accumulation of food reserves, the acquisition of desiccation tolerance, and the induction of seed dormancy.

ABA will reverse the GA induction of an increase in free calcium ions and alpha-amylase in barley aleurone prototplasts. A process which initiates the germination response in barley and other grains.

Signal transduction in animal systems for epidermal growth factor (EGF) follow this simplified scheme: Receptor Kinase → Ras→ Raf→MEK→ MAPK (mitogen-activated protein kinase). Plants, like animals, also balance the processes of the cell cycle and cell differentiation, and they show many parallels to signal transduction in animals.

Ethylene induces the sequential phosphorylation of several proteins that have counterparts in animal cells. The ethylene receptor cannot be directly linked to animal cell receptor kinases like EGF receptors, however, other of the cellular responses to ethylene seem to depend on a number of orthologs of animal signal transducing proteins. Kinases of the Raf, MEK, and MAPK types have all been directly implicated in ethylene signaling, although as negative regulators.

Chapter 5

Seed Dormancy

Seed dormancy is defined as a state in which seeds are prevented from germinating even under environmental conditions normally favorable for germination. These conditions are a complex combination of water, light, temperature, gasses, mechanical restrictions, seed coats, and hormone structures.

In the sprouting industry, dormancy is commonly referred to as a seed producing plants ability to withstand winter freezes. The more dormant or "Winter Hardy", a seed producing plant is, the further north (in northern climates) it can grow and still survive the winter to come up the following season. Often Canadian alfalfa will carry a rating of 2 while Australian will carry a rating of 8 or 9.

Below are the different ratings:

1. Very Dormant
2. Dormant
3. Moderately dormant
4. Categories of semi dormant and intermediate dormancy
5. Moderately non-dormant
6. Non-Dormant (moapa 69 for instance)
7. Very Non-Dormant (cuf 101)

As a practical matter for sprout producers, more dormant seed is generally easier to grow without having spoilage problems, especially in the summer. However, it generally has more hard seed, takes longer to sprout, and has a lower yield. It is also generally less expensive.

BIOLOGY OF A PROCESS

The majority of higher plants derive energy for growth and development solely from photosynthesis (autotrophic nutrition). However, each plant undergoes in its ontogenesis a short but important stage of heterotrophic nutrition. The germinating embryo prior appearance of the first photosynthetic structures is fed exclusively by storages accumulated during the seed formation (i.e., proteins, fats, and carbohydrates).

As is well known, the main function of a plant seed is to provide a viable offspring. The storage substances play a key role, by providing nutrition of a seedling during the heterotrophic stages of its development. Seed formation passes several overlapping but rather independent stages. Before the beginning of flowering, an embryonic structures are formed out of apical meristem.

The switch from the vegetative stage of development to the reproductive one is regulated by different mechanisms. Among these mechanisms are inner biological clock, hormonal background, and environmental conditions such as a day length, temperature fluctuations, humidity, etc.

1. At the first stage, a pollination followed by fertilization occur.
2. The second stage is characterized by establishment of a general composition of the future plant. Embryonic tissues (protoderm, procambium, ground meristem) are being differentiated. The axis of development of an embryo is being formed, with apical meristem of a root from one side and apical meristem of a stem - from the other.
3. At the following stage, storage substances are being intensively worked out, these substances will be

necessary at subsequent stages, during seed germination.

4. Seed development is finished by preparation of a seed to the dormancy stage and final maturation at the dormancy stage.
5. Dormancy stage.

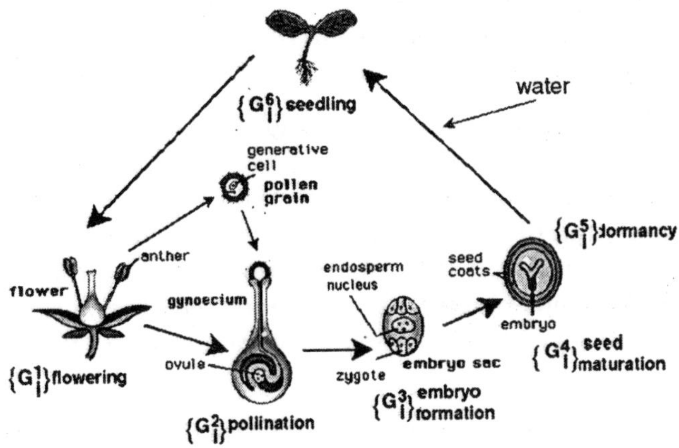

The gene networks conception allows considering a separate phenotypic feature, separate process, stage of development, and ontogenesis in a whole as a result of functioning of gene networks, which are organized hierarchically. Every stage of ontogenesis can be figured out as a result of interaction between gene networks. Elementary events occurring at the cellular level play a considerable role. One and the same gene networks function synchronically in all cells of a certain tissue.

The presence of regulatory contours supports a definite functional state of a gene network, i.e., timely transport of nutrients into developing embryo) or its transformation into another regime of functioning, which may occur under the action of environmental factors (e.g., change of dominant storage protein under nitrogen deficiency). The action of external signals turns one gene network off and turns on the others.

Hormones, Ca++ ions and other low-molecular compounds are molecular carriers of the signals. A gene network with a higher range provides the coordination of this process. A gene network on biosynthesis of storage proteins in plant seeds is under the control of the gene network supporting development of viable seeds. This gene network, in its turn, gets on after a flower's fertilization and is governed by a gene network of a maternal plant having initiated a generative phase of development.

The gene network on storage proteins biosynthesis provides a transition of a seed from the stage of main embryonic structures formation to almost mature state. After accumulating of a sufficient amount of resources, the gene network considered stops its functioning and the genes of storage proteins biosynthesis are repressed. The mechanism terminating an accumulation of storage substances is still unknown.

Seed dormancy could be considered simply as a block to the completion of germination of an intact viable seed under favourable conditions. In our Tansley review 'Seed dormancy and the control of germination' we present an integrated view across the evolution, molecular genetics, physiology, biochemistry, modelling and ecophysiology of the control of seed germination by dormancy in an attempt to draw together these linked, but often separate disciplines.

Definition of seed dormancy: A more sophisticated and experimentally useful definition of dormancy has been proposed by Baskin and Baskin. A dormant seed does not have the capacity to germinate in a specified period of time under any combination of normal physical environmental factors that are otherwise favourable for its germination, i.e. after the seed becomes non-dormant. A completely non-dormant seed has the capacity to germinate over the widest range of normal physical environmental factors possible for the genotype.

Discussion and opinion: Dormancy is a block to germination has evolved differently across species through adaptation to the prevailing environment so that germination occurs when conditions for establishing a new plant generation are likely to be suitable. A diverse range of dormancy mechanisms (blocks) has evolved in keeping with the diversity of climates and habitats in which they operate.

Definitions of dormancy are difficult because dormancy can only be measured by the absence of germination. We can observe completion of germination of a single seed as an all-or-nothing event, whereas dormancy of a single seed can have any value between all (maximum dormancy) and nothing (non-dormancy). Dormancy should not just be associated with the absence of germination, rather it is a characteristic of the seed that determines the conditions required for germination.

When dormancy is considered in this way, any environmental cue that alters the conditions required for germination is by definition altering dormancy. Also by extension, when the seed no longer requires specific environmental cues it is non-dormant. Dormancy is a seed characteristic which defines the conditions required for germination and therefore any cue that widens the environmental requirements for germination should be regarded as a dormancy release factor.

A wide range of factors can therefore alter (physiological) seed dormancy, e.g. temperature, light, nitrate or naturally occurring chemical signals in leachate from litter that covers the seeds in their habitat. However, there is an important distinction in the seeds response to these factors.

1. There are factors that are related to slow seasonal change. These factors (e.g. temperature) are integrated over time to alter the depth of dormancy, and the sensitivity to other factors.
2. There are other factors that indicate in a more immediate way that conditions are suitable for germination (e.g. light), which could be considered to terminate dormancy and therefore induce germination.

Each of these factors therefore remove successive blocks to germination, but this process usually needs to be carried out in a set order for it to work, i.e. in the process described light must come last to be effective. In summary, seed dormancy is an innate seed property, which defines the environmental conditions that must be met before the seed can germinate. The intrinsic molecular mechanisms that determine dormancy have an embryo and/or a coat component. However, dormancy as a "whole seed"-characteristic controls germination and a classification system for seed dormancy has been proposed.

Primary *versus* secondary seed dormancy: Freshly harvested mature water-permeable dormant seeds are said to have primary dormancy, which has been induced with the involvement of abscisic acid (ABA) during seed maturation on the mother plant.

Subsequent dormancy release "in the field", following dispersal, may involve the same factors that are commonly used "in the lab": either after-ripening in the relatively dry state, or dormancy-release treatments in the imbibed state. These imbibed seed treatments include chilling (cold stratification), warm stratification, light, gibberellins (GA) and other hormones, smoke substances like butenolide and compounds like nitric oxide.

In contrast to primary dormancy, secondary dormancy can be induced in seeds with non-deep physiological dormancy after seed dispersal, and is often associated with annual dormancy cycles in the seed bank.

A classification system for seed dormancy: The 'whole-seed view'

Marianna G. Nikolaeva devised a dormancy classification system reflecting that dormancy is determined by both morphological and physiological properties of the seed. Based on this scheme, C. Baskin and J. Baskin have proposed a comprehensive classification system which includes five classes of seed dormancy. The system is hierarchical with these five classes further divided into levels and types Baskin and Baskin.

Physiological Dormancy (PD)

PD is the most abundant form and is found in seeds of gymnosperms and all major angiosperm clades. It is the most prevalent dormancy form in temperate seed banks and the most abundant dormancy class "in the field". PD is also the major form of dormancy in most seed model species "in the lab", including *Arabidopsis thaliana, Helianthus annuus, Lactuca sativa, Lycopersicon esculentum, Nicotiana* spp., *Avena fatua,* and several cereals.

PD Deep

Embryos excised from PD-deep seeds either don't grow or will produce abnormal seedlings. GA treatment does not break their dormancy. Ca. 3-4 months of cold (subtype a) or warm (subtype b) stratification are required before germination can take place.

PD Intermediate

Embryos excised from PD-intermediate seeds produce normal seedlings. GA promotes germination in some (but not all) species. Seeds require 2-3 months of cold stratification. Dry storage (after-ripening) can shorten the cold stratification period. Example: *Acer pseudoplatanus* (PD intermediate).

PD NON-DEEP

The great majority of seeds have non-deep PD. Embryos excised from these seeds produce normal seedlings. GA treatment can break this dormancy and depending on species dormancy can also be broken by scarification, after-ripening in dry storage, and cold (0-10 °C) or warm (>15 °C) stratification. Based on patterns of change in physiological responses to temperature five types of non-deep PD can be distinguished.

Most seeds belong to type 1 or 2, in which the temperature range at which seed germination can occur increases gradually during the progression of non-deep dormancy release from low to higher (type 1, e.g. *Arabidopsis thaliana*) or from high to lower temperature (type 2, e.g. *Helianthus annuus*). In addition, the sensitivity of the seeds to light and GA increases as non-deep PD is progressively released. The molecular mechanisms of Arabidopsis seed dormancy cycling have been investigated by transcriptome analyses.

Morphological Dormancy (MD)

MD is evident in seeds with embryos that are underdeveloped (in terms of size), but differentiated (e.g. into cotyledons and hypocotyl-radical). These embryos are not (physiologically) dormant, but simply need time to grow and germinate. This group does not include seeds with undifferentiated embryos. Example: *Apium graveolens* (Apiaceae).

Morphophysiological Dormancy (MPD)

MPD is also evident in seeds with underdeveloped (in terms of size) embryos, but in addition they have a physiological component to their dormancy. These seeds therefore require a dormancy-breaking treatment, e.g. a defined combination of warm and/or cold stratification which in some cases can be replaced by GA application.

In MPD-seeds embryo growth/emergence requires a considerably longer period of time than in MD-seeds. Seeds with undifferentiated embryos like the Orchidaceae also have a morphological and a physiological component of dormancy, but they are not considered in this classification scheme. Examples for MPD: *Trollius* (Ranunculaceae), *Fraxinus excelsior* (Oleaceae). There are eight known levels of MPD, based on the protocol for seed dormancy break and germination.

Physical Dormancy (PY)

PY is caused by one or more water-impermeable layers of palisade cells in the seed or fruit coat. In seeds PY-seeds, prevention of water uptake develops during maturation drying and the covering layer(s) control water movement (often associated with hardseededness). Seeds will remain dormant until some factor(s) render the covering layer(s) permeable to water.

In nature, these factors include high temperatures, widely fluctuating temperatures, fire, drying, freezing/thawing and passage through the digestive tracts of animals. In seed technology, mechanical or chemical scarification can break PY dormancy. Once PY is broken, i.e. the seed or fruit coat becomes permeable to water, the seeds can germinate over a wide range of ambient conditions.

Unlike PD-seeds, which may re-enter (secondary) dormancy after primary dormancy is broken, once the coat of PY-seeds becomes permeable it generally cannot revert to complete impermeability. Thus, the timing of dormancy break in nature is a more critical event in the life cycle of plants with PY, than it is in those with PD. The mechanism for PY-break must therefore be fine-tuned to the environment.

Specialized Structures for Water-impermeability

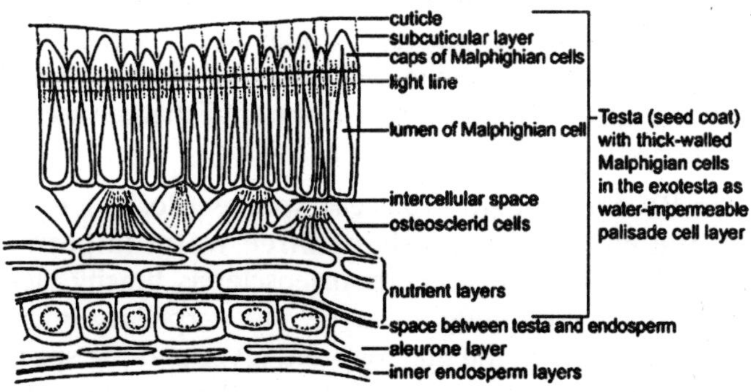

PY is associated with the main mechanical covering layer(s) of the seed and/or fruit coat. In most hardseeded species a palisade of radially elongated cells prevents water uptake. This palisade layer(s) prevent water uptake by their physical arrangement as a tissue and the chemical coatings/ impregnates of the cells (heavy ligninfication, suberin-cutin matrix, waxes). In several cases specialized structures are associated with the control of water-impermeablility.

Water-impermeability of PY-dormant legume seeds is due to a palisade layer of Malphighian cells in the exotesta. The the lens (strophiole, not shown) is the spezialized structure localized in the hilum region that controls water-impermeability.

In PY-dormant Fabaceae species (Rosids) the lens (strophiole) is the spezialized structure localized in the hilum region that controls water-impermeability. The hardseededness and water-impermeability of many legume seeds is due to a palisade epidermal layer of thick-walled Malpighian cells in the outer testa (exotesta). PY-dormancy of legume seeds is released if the lens is disrupted and a 'water-gap' allows water uptake.

In most legume seeds stress, e.g. by heating, causes disruption (pulling apart) of the relatively thin-walled cells of the lens. In some legume species 'pop-up lenses' are evident. There is strong evidence that the lens is the weak point in the palisade layer of PY-dormant Fabaceae seeds. It is however not the only place of water entry: The lens functions as an environmental 'signal detector', it may act as regulator of the rate of water entry into the seed, thereby affecting seed and seedling vigor. Examples for PY-dormant Fabaceae seeds with lens are *Melilotus* and *Trigonella* species.

In PY-dormant Ancardiaceae species (Rosids), e.g. *Rhus* spp., the endocarp consists of three water-impermeable palisade layers (macrosclereids, osteosclereis, brachysclereids) and the outer crystalliferous layer. The carpellary micropyle is a discontinuity in the macroslereid layer and the specilized structure that controls water-impermeability. No discontinuity occurs in the osteosclereid and brachysclereïd layers, and these layers block water entrance to the carpellary micopyle in dormant seeds.

Dormancy release by heating causes formation of a blister near the carpellary micropyle via uplifting of the osteosclereid and brachysclereid layers. During blister formation, a slit is formed in these layers, therby creating an opening for water entrance.

Different types of chalazal plugs/caps are the spezialized structures that control water-impermeability of other seeds, e.g. in PY-dormant species of the Malvales clade (Rosids) families Bixaceae, Cistaceae, and Malvaceae. A suberized 'stopper' formed from the nucellus is the spezialized structure that controls water-impermeability of the Geraniaceae (Rosids). An imbibition lid is the spezialized structure that controls water-impermeability of the Cannaceae (Monocots), the only Monocot family with PY species.

Combinational Dormancy (PY+PD)

Combinational dormancy, PY+PD, is evident in seeds with water-impermeable coats (PY) combined with physiological embryo dormancy (PD non-deep). Within the Ancardiaceae the genus *Rhus* includes species with PY+PD- dormancy (e.g. *Rhus aromatica*) and species with only PY-dormancy (e.g. *Rhus glabra*). Embryos of freshly harvested mature PY+PD-dormant seeds of winter annuals, e.g. *Geranium* spp. (Geraniaceae, Rosids) and *Trifolium* spp. (Fabaceae, Rosids), have some conditional dormancy (PD non-deep).

This PD is released during after-ripening storage, even of the seed coat remains water impermeable (PY). Embryos of PY+PD-dormant *Ceanothus* spp. (Rhamnaceae, Rosids) are more deeply dormant (but still PD non-deep). In this case PY is released first and the subsequent water uptake during imbibition is followed by the release of the embryo dormancy (PD non-deep) during a few weeks of cold stratification. Thus, release from PY and PD of PY+PD-dormant seeds seems to be by independend events and the timely order can be species-specific.

EVOLUTION OF SEED STRUCTURE AND SEED DORMANCY

MD and MPD are present in B-type seeds, but can also be typical of some specialized species. MD is thought to be the ancestral dormancy type among seed plants and is the most primitive dormancy class. The dispersal of seeds with an underdeveloped embryo that need time to grow might have evolved as an ancient strategy to disperse germination over time.

MD and MPD are not only typical for primitive angiosperms, but also for primitive gymnosperms like the Zamiaceae, Ginkgoaceae, Podocarpaceae and Taxaceae. Evolution of larger embryo size in MD seeds resulted in non-dormant (ND) seeds. Concurrently, the gain of physiological dormancy mechanism(s) led from seeds with MD to seeds with MPD, which upon gain in embryo size led to PD seeds.

PD is the most phylogenetically widespread dormancy class. This class of dormancy is distributed over the entire phylogenetic tree from gymnosperms and the basal angiosperms to the higher core eudicot Rosid clade. As ND is also distributed over the entire phylogenetic tree it has been proposed that gain and loss of PD occurred at several times and at several levels of seed evolution.

The most phylogenetically restricted and derived dormancy classes are PY and PY+PD. The occurrence of an impermeable seed or fruit coat combined with a ND embryo (PY) or a PD embryo (PY+PD) is probably the adaptation to specialised life strategies or habitats. PY and PY+PD are the only dormancy classes not found in gymnosperms, and PY is only evident in one Monocot family (Cannaceae).

A conceptual model for the evolution of PY and PY+PD within the Ancardiaceae has been proposed. Within this family the genus *Rhus* includes species with PY+PD- dormancy and species with only PY-dormancy. Anatomical evidence for PY of extinct *Rhus* spp. suggests that PY had evolved in the Ancardiaceae by 43 Ma in the Middle Eocene. The model begins in pre-Eocene times with an ancestral species that has large, non-dormant (ND), recalcitrant seeds.

By the middle Eocene, a derived species with relatively small, orthodox seeds and a water-impermeable covering layer had evolved, i.e. a PY-dormant seed evolved. By the Oligocene, PD had been added to the seed dormancy mechanism, i.e. a

PY-dormant seed evolved. Climatic drying (Eocene), followed by climatic cooling (Eocene-Oligocene transition), are proposed as primary selective agents in the development of PY.

An evolutionary connection between PY and recalcitrance is suggested by the relatively high concentration of these two states in the Rosid clade. Phylogenetic data and fossil evidence seem to support the PY —> PY+PD evolutionary sequence in Ancardiaceae, which also may have occured in Fabaceae.

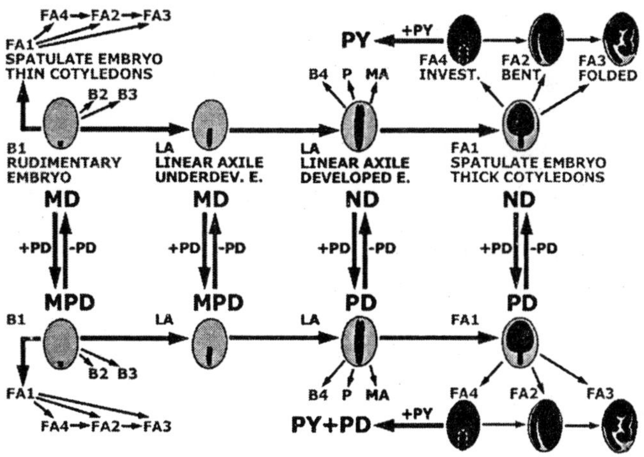

Summary of general trends in angiosperm seed evolution with respect to seed type and dormancy class.

Seeds with small embryo and abundant endosperm (B1 seed type) are primitive and evolution shows a general trend (indicated by arrows) towards mature seeds with little or no endosperm in which the embryo occupies most of the seed (through LA to FA seed types). Gain or loss of PD to each of these seed types (B1, LA and FA1) alters the dormancy class as indicated.

Similar changes can also occur in the other seed types. Addition of PY to FA4-type ND or PD seeds is indicated as a side branch occurring in specialized species. This addition can be reversed and PY can also be added to other seed types.

Embryo Dormancy and Coat Dormancy: The components of Physiological Dormancy

The hierarchical classification system by Baskin and Baskin accurately reflects the "whole-seed" ecologists view of the control of germination by dormancy. At the molecular level very little is known about MD, MPD, PY, PY+PD and deep PD. Recent work has provided insight into molecular mechanisms of non-deep PD.

Together with earlier physiological work it is clear that the intrinsic molecular mechanisms determining dormancy can have an embryo and/or a coat component. The terms embryo and coat dormancy are used to distinguish between these two mechanisms, but it should be emphasized that they are not used as part of a classification system for "whole-seeds" sensu Baskin and Baskin.

Embryo dormancy and coat dormancy are components of PD; their sum and interaction determines the degree of "whole-seed" PD. Embryo dormancy is characterized by a block that inhibits extension growth and therefore excised embryos do not grow. Coat dormancy, is characterized by a block that is conferred by the covering layers. Non-dormant embryos excised from coat-dormant seeds will therefore extend and grow.

"Coat" is used in a loose sense and can be any embryo covering structure, e.g. testa, endosperm and/or pericarp and terms like testa dormancy and endosperm dormancy can be used to specify these. Mechanical resistance from combined testa and endosperm dormancy, which is greater than the embryo growth potential opposing it, appears to be the cause of non-deep PD in seed model systems like *A. thaliana* and Solanaceae species. When embryo and coat dormancy are considered as components of non-deep PD the influence of ABA and GA on dormancy and the control of germination can be summarised as follows:

Embryo Dormancy

Embryo dormancy (embryo component of non-deep PD): A dormant embryo is characterized by a high ABA/GA ratio, high ABA sensitivity and low GA sensitivity. Embryo dormancy release involves remodelling of hormone biosynthesis and degradation towards a low ABA/GA ratio, a decrease in ABA sensitivity and an increase in GA sensitivity.

Thus ABA dominates the embryo dormancy programme, and GA the embryo germination programme. A non-dormant embryo is characterized by increased growth potential, the ability for cell extension growth and the ability to induce the release of coat dormancy.

Coat Dormancy

Coat dormancy: The combination of an embryo with low growth potential and mechanical constraint from the seed covering layers can result in dormancy. GA can release this coat dormancy by increasing the embryo growth potential and/or by reducing the mechanical constraint.

There are two forms which are distinguished based on whether the covering layer consists of dead (e.g. outer testa) or living (e.g. endosperm) tissue:

1. Testa dormancy: ABA during seed development determines the subsequent GA requirement for germination since ABA influences the testa characteristics (e.g. thickness) and GA the embryo growth potential.
2. Endosperm dormancy: Endosperm weakening can be either part of the coat dormancy release or part of the germination programme. Since the endosperm is in most cases living tissue it can actively participate in regulating embryo constraint by influencing both the ABA/GA ratio and sensitivity to these hormones. GA acts by increasing the embryo growth potential and by promoting endosperm weakening which is achieved though ABA-independent and ABA-inhibited mechanisms.

Induction, maintenance and release of physiological dormancy by plant hormones and environmental signals.

There is considerable evidence that abscisic acid (ABA) is an important positive regulator of both the induction of dormancy and the maintenance of the dormant state in imbibed seeds following shedding.

- ABA-deficiency during seed development is associated with absence of primary dormancy in the mature seed, whereas over-expression of ABA biosynthesis genes can increase seed ABA content and enhance seed dormancy or delay germination.
- ABA produced by the seed itself during seed development can impose a lasting dormancy, whereas, maternal ABA or ABA application during seed development both fail to induce a lasting seed dormancy, but has other functions.
- In *Arabidopsis thaliana,* the members of the AtNCED gene family encode enzymes catalyzing the key regulatory step in ABA biosynthesis. ABA synthesis in the embryo and the endosperm both seem to contribute to the induction of seed dormancy.
- High ABA contents are present in the imbibed seeds of the strongly dormant *A. thaliana* ecotype Cape Verde Island (Cvi) and decrease as dormancy is lost. A recent transcriptome analysis with this ecotype strongly supports the view that increased ABA biosynthesis is associated with the dormant state.
- *De novo* ABA biosynthesis during imbibition of dormant, but not non-dormant, seeds has been demonstrated in the *A. thaliana* ecotype Cvi, as well as in other species including *Nicotiana plumbaginifolia, Helianthus annuus,* and *Hordeum vulgare.* This de novo ABA biosynthesis has been interpreted as a mechanism for dormancy maintenance.
- Work with the strongly dormant *A. thaliana* ecotype Cvi shows that dormancy may depend on an intrinsic balance of gibberellin (GA)- and ABA-biosynthesis and catabolism, which will determine the dominance of either of the hormones. While PD release of Cvi

seeds occurs effectively by after-ripening, stratification or inhibition of ABA biosynthesis, the addition of GA appears less effective. GA treatment of dormant Cvi seeds caused a transient increase in ABA levels, suggesting that in dormant seeds a feedback mechanism exists that maintains a high ABA/GA ratio. Thus, the net result of the dormant state is characterized by increased ABA biosynthesis and GA degradation. It appears to be the ABA/GA ratio, and not the absolute hormone contents, that controls germination.

Model for the Regulation of Dormancy and Germination by ABA and GA in Response to the Environment

According to this model ambient environmental factors (e.g. temperature) affect the ABA/GA balance and the sensitivity to these hormones. ABA synthesis and signalling (GA catabolism) dominates the dormant state, whereas, GA synthesis and signalling (ABA catabolism) dominates the transition to germination.

The complex interplay between hormone synthesis, degradation and sensitivities in response to ambient environmental conditions can result in dormancy cycling. Change in the depth of dormancy alters the requirements for germination (sensitivity to the germination environment); when these overlap with changing ambient conditions, germination will proceed to completion.

VARIATION OF THE SEED DORMANCY TRAIT IN *ARABIDOPSIS* AND *NICOTIANA*

Arabidopsis Thaliana

Non-dormant seeds of *Arabidopsis thaliana* germinate in the presence of water, oxygen, light and a suitable temperature, but dormant *Arabidopsis* seeds do not germinate at these favorable ambient conditions. *Arabidopsis* is native to Europe and central Asia. Flowering time and seed dormancy are key traits that determine the timing and lenght of the *Arabidopsis* natural life cycle.

For both traits extensive natural variation is present among the different *Arabidopsis* accessions, which determine the variation observed in life cycles and the climatic adaptations. In Europe, Arabidopsis flowers most abundantly in spring and early summer, and mature seeds are dispersed from May to July. These seeds exhibit a certain degree of conditional dormancy (PD non-deep), most of them germinate in the fall and survive winter as rosettes and therefore are winter annuals.

Under greenhouse conditions these Northern European winter annual accessions flower late, but are very responsive to vernalization (cold treatment), which under natural conditions synchronizes flowering in springtime. In general, most accessions collected from nature in Northern Europe are late flowering are typical winter annuals; Southern populations are of both types, summer and winter annuals.

Seed dormancy is a quantitative trait, determined by several genes with an additional strong impact of the environment. The analysis of natural alleclic variation at seed dormancy loci of *Arabidopsis thaliana* was performed by exploiting these differences of accessions in the degree of dormancy.

The laboratory strains Columbia (Col) and Landsberg *erecta* (L*er*) have low dormancy, whereas the Cape Verde Island (Cvi) and C24 accessions have deep dormancy. Recombinant inbred lines (RILs) / near isogenic lines (NILs) between Cvi and L*er* were used in a quantitative trait loci (QTL) approach and identified several loci, named *"Delay Of Germination 1* to *12"*, that determine *Arabidopsis* seed dormancy. *DOG1*, for which the Cvi allele increased the level of seed dormancy, explains 12% of the variance observed in seed dormancy in these RIL populations.

The *DOG1* gene has been cloned and is a member of a small gene family of unknown molecular function. The functional allelic variation present in Arabidopsis is caused by polymorphisms in the cis-regulatory region of the *DOG1* gene resulting in considerable expression differences between

DOG1 allels of the accessions analyzed. Alternative splicing leading to DOG1 proteins that differ in the C-terminal domain also appears to be important for dormancy.

Seed dormancy and germination of *Arabidopsis* are controlled by phytohormones and an interaction between the growth potential of the radicle and the mechanical resistance of the seed covering layers. Two-step seed germination of *Arabidopsis* is visible as (1) testa rupture and (2) endosperm rupture. Only the endosperm rupture step is inhibited by ABA, but ABA during seed maturation is involved in determining the degree of Arabidopsis seed dormancy.

Germination of dormant Arabidopsis seeds is blocked prior to testa rupture, and GA is required for its induction. Seed dormancy of mutants affected in ABA and GA levels and sensitivity is altered: ABA-deficiency (*aba1* mutant) and ABA-overproduction (*cyp707a* mutants) lead to non-dormancy and increased dormancy, respectively. The ABA-insensitive mutant *abi3* is non-dormant.

GA-deficiency (*ga1* mutant) leads to increased dormancy. Mutations in DELLA-type repressors mediating of GA-responses in seeds lead either to enhanced (gain-of-function) or to reduced (loss-of-function) dormancy.

Control of germination by the endosperm - *Arabidopsis* seed dormancy is primarily determined by the endosperm: A publication demonstrates that the Arabidopsis endosperm layer is sufficient and necessary for seed dormancy. Embryos from dormant seeds had a lesser growth potential than those from nondormant seeds.

However, the endosperm layer was found to be the primary determinant of seed dormancy. Arabidopsis endosperm (aleurone) cells responded to nitric oxide (NO), GA, and ABA, with NO being upstream of GA in a signaling pathway that leads to vacuolation (GA-induced, ABA-inhibited) of protein storage vacuoles. Dormancy release and the GA signal lead to Arabidopsis endosperm rupture which involves these microscopic changes that are typical for endosperm weakening.

Control of germination by the seed coat - *Arabidopsis* testa mutants: Completion of germination in seeds with PD-type coat dormancy requires that either the embryo growth potential increases to overcome the mechanical constraint, and/ or the mechanical constraint associated with the seed covering layer(s) is reduced.

Two forms of mechanical constraints and release mechanisms can be distinguished:

1. Living seed covering layers where a regulated tissue weakening occurs prior to germination and the tissue itself can produce enzymes for this process, e.g. of endosperm or inner testa;
2. Mostly dead seed covering layers where predetermined breaking points facilitate tissue rips prior to germination, e.g. outer testa or pericarp.

The testa is maternal tissue and the reduced seed dormancy phenotype is inherited maternally. A series of *Arabidopsis* testa mutants show reduced dormancy that is caused by alterations of the testa characteristics. Examples are *tt* (*transparent testa*), *ttg* (*transparent testa glabra*), and *ats* (*aberrant testa shape*). The testa mutants highlight the importance of the testa structure as a constraint to radicle emergence. The GA requirement for *Arabidopsis* seed germination is determined by testa characteristics, embryonic growth potential and by embryonic ABA.

Arabidopsis seed dormancy genes: These genes are represented by *Arabidopsis* mutants affected in seed dormancy and include hormone-related genes, and other genes that affect dormancy via the testa characteristics or by different mechanisms, e.g. *fus3* (*fusca3*), *lec1* and *lec2* (*leafy cotyledons*), *dag1* and *dag2* (*DOF affecting germination*, DOF-type transcription factors), *rdo1* to *rdo4* (*reduced dormancy*), *cts* (*comatose*), and *DOG1* (*delay of germination*).

Nicotiana spp. - Tobacco

Nicotiana spp. seed dormancy and ABA: The onset of dormancy in *Nicotiana tabacum* is correlated with a peak in

ABA content at approximately 15-20 days after pollination (DAP), and a rapid decline in ABA content follows during further seed maturation. Tobacco seeds harvested after DAP 25 are dark-brown, the embryos are mature and have white cotyledons, the products of maturation-specific genes have accumulated, dormancy has been established, desiccation tolerance has been acquired, and the moisture content is low.

Seed dormancy is not established in transgenic tobacco expressing an anti-ABA antibody that causes deficiency in free ABA, resulting in precocious germination, embryos with green cotyledons, reduced accumulation of storage proteins, and desiccation intolerance. *Nicotiana* species differ in the degree of seed dormancy (PD non-deep): *N. tabacum* and *N. sylvestris* have a low level of dormancy, while *N. plumbaginifolia* has a somewhat higher degree of dormancy.

The high degree of seed dormancy of the post-fire annual tobacco, *Nicotiana attenuata,* can be induced by ABA and terpenes and smoke-exposure, light and GA are required for dormancy release.

The wilty mutants *aba1* and *aba2* of *Nicotiana plumbaginifolia* have reduced ABA contents and exhibit precocious germination and reduced primary dormancy. ABA-deficiency of the *aba2* mutant is due to a mutation in the *ABA2* gene, encoding zeaxanthin epoxidase, a key step in ABA biosynthesis. Antisense- and sense-ABA2 transformation of *N. plumbaginifolia* resulted in decreased and increased ABA biosynthesis and seed dormancy, respectively.

Experiments suggested that, as in Arabidopsis and tomato, also in *Nicotiana* species only ABA produced by the seed itself, but not maternal ABA, is necessary to impose a lasting dormancy. Tobacco dormancy can be released during after-ripening, i.e. a period of dry storage of freshly harvested, mature seeds. A further decline in ABA content, decreased sensitivity to ABA and increased sensitivity to GA are involved in the after-ripening-mediated transition from the dormant to the non-dormant state of many species.

The work of Grappin et al. demonstrated this for *N. plumbaginifolia* and showed in addition *de novo* ABA biosynthesis in imbibed fresh (dormant), but not after-ripened (non-dormant) seeds. Promotion of ABA-delayed seed germination of *N. plumbaginifolia* by light or GA involves stimulation of ABA degradation and inhibition of ABA synthesis

Tobacco photodormancy: Light is required for at least two aspects of tobacco seed germination. First, in photodormant tobacco seeds germination in darkness is blocked at a step before testa rupture and neither neither testa nor endosperm rupture occur even after several weeks of dark-imbibition. Brief treatment of imbibed photodormant seeds with red light activates the phytochrome signal transduction pathway resulting in the release of photodormancy and the promotion of germination.

Genetic and physiological experiments suggest that tobacco photodormancy is mainly under maternal control. Gibberellin (GA) can substitute for the red-light trigger needed to release photodormancy and to induce testa rupture and subsequent endosperm rupture of tobacco seeds imbibed in the dark.

Red light has been shown to up-regulate the biosynthesis of bioactive GA in germinating seeds of lettuce and Arabidopsis. Fresh tobacco seeds are photodormant and after-ripening contributes to the release of photodormancy. This effect varies greatly for different seed batches as reported for several tobacco cultivars. The GA requirements for photodormancy release of fresh and completely photodormant after-ripened seed batches are equal.

Non-photodormant tobacco seeds have lost the GA requirement for dark-germination, which could be due to increased GA sensitivity and/or increased endogenous GA levels. Light is also required for a second aspect of tobacco seed germination, it promotes the speed of endosperm rupture of non-photodormant tobacco seeds. GA is also not only involved in inducing dark-germination of photodormant tobacco seeds, it also promotes ABA-delayed endosperm rupture of dark-imbibed non-photodormant seeds, and osmoticum-delayed testa and endosperm rupture of light-imbibed seeds.

Finally, endogenous ethylene and brassinosteroids also promote the germination of non-photodormant tobacco seeds and counteract the inhibitory effects of ABA on endosperm rupture, but ethylene and brassinosteroids do not release tobacco photodormancy.

Types of Seed Dormancy and the Roles of Environmental Factors

During seed maturation, the embryo enters a quiescent phase in response to desiccation. Seed germination can be defined as the resumption of growth of the embryo of the mature seed; it depends on the same environmental conditions as vegetative growth does. Water and oxygen must be available, the temperature must be suitable, and there must be no inhibitory substances present.

However, in many cases a viable (living) seed will not germinate even though all the necessary environmental conditions for growth are satisfied. This phenomenon is termed seed dormancy. Seed dormancy introduces a temporal delay in the germination process that provides additional time for seed dispersal over greater geographical distances. It also maximizes seedling survival by preventing germination under unfavorable conditions.

Two types of seed dormancy have been recognized, coat-imposed dormancy and embryo dormancy. Coat-imposed dormancy is dormancy imposed on the embryo by the seed coat and other enclosing tissues, such as endosperm, pericarp, or extrafloral organs. The embryos of such seeds will germinate readily in the presence of water and oxygen once the seed coat and other surrounding tissues are either removed or damaged. There are five basic mechanisms of coat-imposed dormancy:

1. *Prevention of water uptake:* Prevention of water uptake by the seed coat is a common cause of seed dormancy

in families found in arid and semiarid regions, especially among legumes, such as clover (*Trifolium* spp.) and alfalfa (*Medicago* spp.). Waxy cuticles, suberized layers, and lignified sclereids all combine to restrict the penetration of water into the seed.

2. *Mechanical Constraint:* The first visible sign of germination is typically the radicle breaking through the seed coat. In some cases, however, the seed coat may be too rigid for the radicle to penetrate. Nuts with hard, lignified shells are examples of dormancy caused by mechanical constraint. Such shells must be broken by biotic or environmental forces for the seed to germinate. Even nonlignified tissues, such as the endosperm of lettuce seeds, can suppress expansion of the embryo. For the seeds to germinate, the endosperm cell walls must be weakened by the production of cell wall–degrading enzymes.
3. *Interference with gas exchange:* Some seed coats are considerably less permeable to oxygen than an equivalent thickness of water is— e.g., less by a factor of about 10^4 in seeds of cocklebur (*Xanthium pennsylvanicum*). This lowered permeability to oxygen suggests that the seed coat inhibits germination by limiting the oxygen supply to the embryo. In support of this idea, investigators can break the dormancy of such seeds either by making a small hole in the coat with a pin (without weakening the coat mechanically), or by treating the coat with concentrated oxygen. However, other studies suggest that the oxygen consumption of the embryos from such seeds is considerably less than the amount of oxygen able to penetrate the seed coats under normal aerobic conditions. Thus the role of oxygen impermeability in seed coat dormancy remains unresolved.
4. *Retention of inhibitors.* The seed coat may prevent the escape of inhibitors from the seed. Growth inhibitors readily diffuse out of isolated *Xanthium* embryos, but not from intact seeds.

5. *Inhibitor production.* Seed coats and pericarps may contain relatively high concentrations of growth inhibitors that can suppress germination of the embryo. ABA is a common germination inhibitor present in these maternal tissues. In certain cases where repeated washing (leaching) removes dormancy, the effect is thought to be due to the loss of such inhibitory compounds.

The second type of seed dormancy is embryo dormancy, which refers to a dormancy that is inherent in the embryo and is not due to any influence of the seed coat or other surrounding tissues. In some cases, embryo dormancy can be relieved by amputation of the cotyledons. Species in which the cotyledons exert an inhibitory effect include European hazel (*Corylus avellana*) and European ash (*Fraxinus excelsior*). A fascinating demonstration of the cotyledon's ability to inhibit growth is found in species (e.g., peach) in which the isolated dormant embryos germinate but grow extremely slowly to form a dwarf plant. If the cotyledons are removed at an early stage of development, however, the plant abruptly shifts to normal growth.

Embryo dormancy is thought to be due to the presence of inhibitors, especially ABA, as well as the absence of growth promoters, such as GA (gibberellic acid). The loss of embryo dormancy is often associated with a sharp drop in the ratio of ABA to GA.

Seed Dormancy May be Primary or Secondary

Different types of seed dormancy also can be distinguished on the basis of the timing of dormancy onset rather than the cause of the dormancy.

- Seeds that are released from the plant in a dormant state are said to exhibit primary dormancy.
- Seeds that are released from the plant in a nondormant state but which become dormant if the conditions for germination are unfavorable exhibit secondary dormancy. Seeds of *Avena sativa* (oat) can

become dormant in the presence of temperatures higher than the maximum for germination, whereas seeds of *Phacelia dubia* (small-flower scorpionweed) become dormant at temperatures below the minimum for germination. The mechanisms of secondary dormancy are poorly understood.

Environmental Factors Control the Release from Seed Dormancy

Various external factors release the seed from dormancy, and dormant seeds typically respond to more than one condition. Many seeds lose their dormancy when their moisture content is reduced to a certain level by drying. This method of breaking seed dormancy is called afterripening, and is usually performed in a special drying oven. On the other hand, if the seed becomes too dry (5% water content or less), the effectiveness of afterripening is diminished.

Another factor that can release seeds from dormancy is low temperature, or chilling. Many seeds require a period of cold (0 to 10°C) while in a fully hydrated (imbibed) state in order to germinate. In temperate-zone species, this requirement is of obvious survival value, since such seeds will germinate not in the fall, but only in the following spring.

Chilling seeds to break their dormancy is a time-honoured practice in horticulture and forestry and traditionally has been referred to as *stratification*. This term is derived from the old agricultural practice of allowing seeds with a chilling requirement to overwinter outdoors in layered mounds of moist soil. Today the seeds are simply stored in a refrigerator.

The third external factor that plays an important role in breaking seed dormancy is light. Many seeds have a light requirement for germination, which may involve only a brief exposure, as in the case of lettuce, an intermittent treatment (e.g., succulents of the genus *Kalanchoe*), or even a specific photoperiod involving short or long days.

Interestingly, all light-requiring seeds exhibit seed coat dormancy, and removal of the outer tissues of the seed allows the embryo to germinate in the absence of light. The effect that light has on the embryo is thus to enable the radicle to penetrate the seed coat. This penetration often involves some enzymatic weakening of the enclosing tissues.

THE LONGEVITY OF SEEDS

How long can a seed remain dormant and still remain viable? Seed longevity is of practical importance because of ongoing efforts to preserve plant genetic resources for future agricultural crops by setting up seed gene banks. Spectacular claims of dormant seeds that remained viable for thousands of years have been made, but are considered highly controversial.

Extreme examples of ancient seeds that have been reported to retain their viability include submerged lotus seeds found in a 3,000-year-old boat near Tokyo, barley seeds from the 3,000-year-old tomb of King Tutankhamen, and arctic lupine seeds associated with rodent burrows determined to be 14,000 years old. In most cases, however, a scientifically rigorous examination of the data has either disproved or raised serious doubts about the claimed antiquity of the seeds.

The most serious criticism is that the dates of the seeds have been inferred from the age of their immediate surroundings or associated artifacts rather than from direct measurements of the seeds themselves. This question raises the possibility that the seeds are much younger than the site at which they were found. In fact, microscopic examination of authenticated ancient cereal grains from various Middle Eastern sites has shown that cells and tissues of these seeds are no longer intact. Such seeds are clearly inviable.

Although the most sensational claims for seed longevity are almost certainly bogus, seeds of *Canna compacta* apparently can live for at least 600 years. Viable *Canna* seeds were obtained from inside a walnut in a tomb in Argentina. The *Canna* seeds had apparently been inserted into the immature seeds of a growing walnut fruit before the hard outer shell formed.

Once the shell hardened and the nut dried out, the result was a rattle. Native people strung the rattles together to form a necklace. In this case, the seeds had to be at least as old as the walnut shell, and carbon dating of the shell indicated that it was about 600 years old.

Why are some seeds able to survive in a dormant state for hundreds of years, whereas others lose viability in less than 5 years? Almost nothing is known about the mechanisms that determine the longevity of seeds. If these mechanisms could be understood, we might someday be able to greatly increase the seed longevity of agriculturally important species and varieties, thereby enhancing our ability to preserve plant genetic resources for generations to come.

GENETIC MAPPING OF DORMANCY

Complex traits, e.g. dormancy, are affected by many different, often unlinked genes. Consequently, attempts to map them result in continuous, or quantitative, variation within a mapping population. Quantitative trait locus (QTL) mapping is used in these instances to map individual genes affecting the trait. In QTL mapping, inbred lines that differ in the trait of interest are crossed and an F_2 population is produced.

The F_2 population has continuous variation for the trait of interest, and the goal of the analysis is to determine which chromosomal segments contain genes affecting the trait. Using marker loci along each chromosome (loci that have different alleles between the parental inbred lines), the genotypes of entire chromosomal segments in each F_2 individual can be inferred as being homozygous for one or the other parental lines or heterozygous.

If the inheritance of a chromosomal segment is correlated with the trait of interest, then a gene affecting the trait is likely to be between the two marker loci. By analyzing chromosomal segments along each of the chromosomes, the number and locations of genes affecting the quantitative trait can be determined.

Two parental (P) inbred lines, A & B, which differ for a trait of interest (e.g. dormancy) are crossed to produce heterozygote F_1 progeny which have an intermediate degree of dormancy. Only one chromosome pair is depicted for simplicity. The F_1 progeny are crossed to produce a large segregating F_2 population. Because of recombination during F_1 meiosis, the F_2 individuals have varying genetic combinations of the original inbred lines (depicted by shading of chromosomes) and have continuous variation, from high to low, in degree of dormancy.

In order to estimate whether segments of the chromosome are homozygous for one of the parental-lines or heterozygote, the F_2 individuals are scored for marker loci (1-6) along the chromosome. Loci 1 & 2 are identified as homozygous for A-line alleles in individual III, as heterozygous in individuals I & V and as homozygous for B-line alleles in individuals II & IV.

The F_2 individuals are also scored for their dormancy. After scoring many F_2 individuals a correlation between the genotypes at locus 1 & 2 and phenotype is determined and therefore this segment likely contains at least one gene affecting dormancy. The analysis is repeated for each chromosomal segment to identify which ones are likely to contain genes affecting dormancy.

Candidate genes are specific loci, identified by predicted biochemical function or homology to regulatory genes from other species, that are suspected of affecting the trait of interest. Molecular differences between the parental alleles at these loci can be used to determine whether they map to the same region as any of the QTLs, thereby supporting the hypothesis that a given gene contributes to regulation of the trait.

Chapter 6

Seed Treatment

Most seed treatment products are fungicides or insecticides applied to seed before planting. Fungicides are used to control diseases of seeds and seedlings; insecticides are used to control insect pests. Some seed treatment products are sold as combinations of fungicide and insecticide.

Fungicidal seed treatments are used for three reasons:

1. To control soil-borne fungal disease organisms that cause seed rots, damping-off, seedling blights and root rot;
2. To control fungal pathogens that are surface-borne on the seed, such as those that cause covered smuts of barley and oats, bunt of wheat, black point of cereal grains, and seed-borne safflower rust;
3. To control internally seed-borne fungal pathogens such as the loose smut fungi of cereals.

Most fungicidal seed treatments do not control bacterial pathogens and most will not control all types of fungal diseases, so it is important to carefully choose the treatment that provides the best control of the disease organisms present on the seed or potentially present in the soil.

The degree of control will vary with product, rate, environmental conditions and disease organisms present. Some systemic fungicidal seed treatments may also provide protection against early-season infection by leaf diseases.

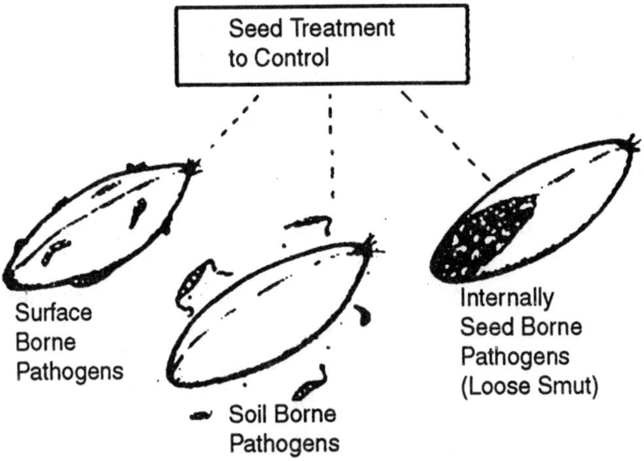

Fig. Reasons for seed treatment.

Fungicide-insecticide combination products or an addition of insecticide for wireworm control should be considered if planting newly opened land or land that has had a history of wireworms. Consult current recommendations for insecticides registered for wireworm control. A fungicide-insecticide combination also may be useful for dry beans. The insecticide used on dry beans should be one that provides control of the seed corn maggot.

SEED TREATMENT APPLICATION

Fungicide seed treatment products come in a variety of formulations and in a variety of packaging sizes and types. Some are registered for use only by commercial applicators using closed application systems, others are readily available for on-farm use as dusts, slurries, water soluble bags, or liquid ready-to-use-formulations.

Whatever the formulation used or application method chosen, some precautions should be taken to assure applicator safety and appropriate seed coverage.

CAUTIONS

Follow label directions when handling seed treatment chemicals. These products are potentially poisonous if mishandled or misused. Extreme caution must be used when handling seed treatment chemicals: some are toxic, others may be irritating. An approved chemical respirator and goggles are recommended even if not specifically required by the fungicide label.

The rate of application prescribed by the label must be used: over treatment may injure the seed and under treatment may not provide good disease control. To apply the correct rate, it is essential to calibrate application equipment carefully and to check calibration frequently. Metering cups of commercial applicators should be cleaned daily to prevent a buildup of chemical that might result in reduced application rates.

An auger which has been used to treat seed cannot be cleaned up sufficiently for use in augering grain for food or feed. Once an auger is used for seed treatment, it should be used only for treatment or augering seed for planting. It should not be used to auger grain used for food or feed.

Treated seed should not be used for food or feed, and treated grain should not contaminate grain delivered to elevators or be placed in bins or in trucks delivering to elevators.

Containers should be triple rinsed with the rinse water added to the treatment mixture. The rinsed containers should be punctured and crushed for disposal in an approved landfill.

CEREAL SEED TREATMENTS

Fungicidal seed treatments help control soil-borne pathogens that cause seed decay, seedling blight and root rot. Control of these diseases may result in better stands, more vigorous seedlings, and increased yields. Protectant fungicides such as captan, maneb, PCNB, thiram, or fludioxonil (Maxim) help control most types of soil-borne pathogens, except for common root rot and take-all.

Protectant fungicides containing captan, maneb, or thiram are sold under various trade names. Fungicidal seed treatments to protect against the soil-borne fungi that cause common root rot and take all will be discussed under barley seed treatment.

Fungicidal seed treatment controls most, but not all, of the seed-borne diseases of small grains. Seed treatment should be used when the seed is contaminated with smut, scab, or black point fungi. Specific recommendations for these diseases are discussed under barley and wheat. No seed treatment is a substitute for good seed.

Barley

Barley has three smuts: covered smut, black semi-loose smut (nigra smut), and loose smut. Covered smut and black semi-loose smut are surface-borne fungal pathogens that infect the emerging seedling. These two smuts can be controlled by various protectant fungicides. Loose smut infects the embryo of the seed before harvest.

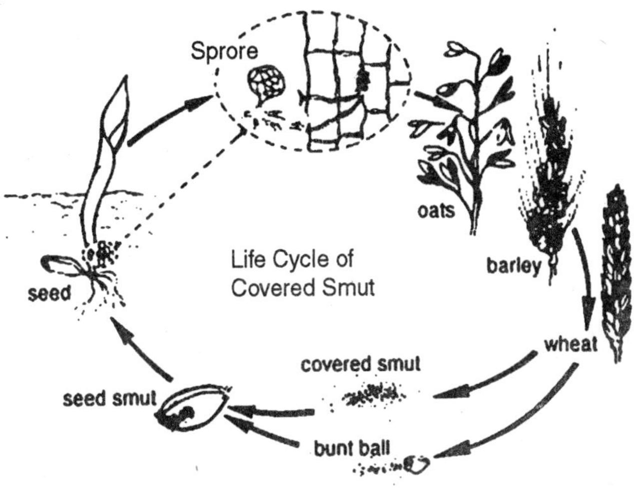

Fig. Life cycle of covered smut and bunt.

Protectant fungicides do not control loose smut—only the systemic fungicides carboxin (Vitavax or Enhance), triadimenol (Baytan) or tebuconazole (Raxil) will control loose smut in barley. These same products also will control the covered and semi-loose smuts of barley.

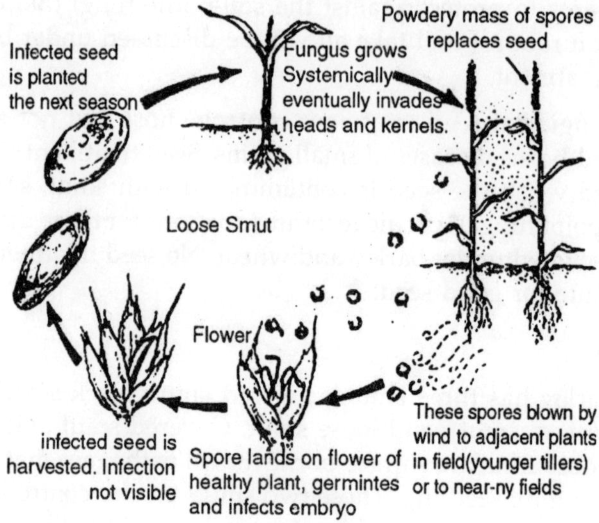

Fig. Life cycle of loose smut.

Covered Smut

Seed of barley varieties susceptible to covered smut should be treated with a protectant fungicide or with carboxin, triadimenol or tebuconazole. At present, all barley varieties recommended for production in North Dakota must be considered susceptible.

Black Semi-loose Smut

Many varieties are susceptible to this smut. It is indistinguishable from loose smut in the field, but it is borne on the seed surface and infects seedlings just as covered smut does. It is not detected by the embryo test, because it does not infect the embryo. It is controlled by seed treatment with a protectant fungicide or with carboxin, triadimenol, or tebuconazole.

Loose Smut

Many of the barley varieties released in the late 1970s and early 1980s were resistant to the races of the loose smut fungus present at that time. In the mid-1980s, a new race of the loose smut fungus was detected in North Dakota.

Seed treatment for loose smut in barley planted for seed production is especially important. Seed samples of susceptible barley varieties should be sent to the North Dakota State Seed Department for an embryo test to determine the percentage of loose smut infection.

Losses from loose smut are about equal to the percent infection; 5 percent loose smut represents a yield loss of about 4.3 percent. If the embryo test shows 2 percent or greater infection, the seed should be treated with carboxin, triadimenol or tebuconazole, or loose smut-free seed should be used for planting.

Common Root Rot

The fungus that causes common root rot is soil-borne and is widespread in North Dakota soils. The common root rot fungus increases under barley or wheat culture, and it often causes mature plant root rot. The fungus also may cause severe root infection in the seedling stage and if splashed to the leaves and head causes spot blotch and black-point infections, respectively.

A number of seed treatment fungicides are now labeled for wheat and barley for suppression of mature plant root rot and also for suppression of seedling blight due to the common root rot fungus. Imazalil is available under several trade names. Triadimenol (Baytan) also is registered for suppression of seedling blight due to common root rot. Tebuconazole (Raxil) is registered for wheat and barley, and difenoconazole (Dividend) is registered only for wheat at the time of publication of this circular.

The conditions in which seed treatment for common root rot would be most beneficial are: where continuous wheat or barley is grown, or if short rotations between these susceptible crops are practiced, and in soils where moisture stress is very likely.

Barley Stripe

This fungus disease is rarely seen in North Dakota but occasionally appears in two-row barley if contaminated seed is planted. The disease causes distinct yellow, then brown stripes that run the entire length of the leaves. The stripes join together and the leaves become split and frayed or shredded in appearance.

The fungus is seed-borne in the hull and seed coat. If seed from a suspected disease source is used, seed treatment with a carboxin + thiram or tebuconazole product provides some control of the barley stripe fungus. The systemic fungicide imazalil provides good control of barley stripe.

CORN

Corn seed is especially susceptible to attack by soil-borne pathogens when sown in cold wet soil, when the seed is in poor condition, when it is mechanically injured, or if it has been stored for two years or more.

Seed treatment will protect against seed rot and reduce the danger of seedling blight. Sweet corn is more susceptible to attack than field corn, but both should be treated. Most field corn is already treated when purchased. A number of protectant and systemic fungicides are registered for control of seedling blights and seed rot in corn.

WHEAT

Bunt or Stinking Smut

The fungus causing this disease adheres to the seed surface and then infects the emerging wheat seedling. The bunt fungus has a fishy odor and imparts the same odor to flour made from bunted kernels. The price of bunted wheat is discounted for this reason. Bunt is not known to occur in North Dakota at the present time, nor has it occurred in recent years. However, it occurs in other wheat-producing states in the Great Plains.

Growers who purchase seed from out of state should use a seed treatment fungicide, since out-of-state seed could be infested with (carrying) bunt. Growers who have their crop custom combined and then save their own seed should use a fungicide, since the combine could be contaminated with bunt spores or bunt balls from states to the south. Many protectant, systemic, or combination products are available to control bunt.

Loose Smut

Loose smut of wheat infects the embryo, as with loose smut of barley. However, the wheat loose smut pathogen does not infect barley and the barley loose smut pathogen does not infect wheat. No embryo test is available for detection of loose smut in wheat.

Counts of loose smut-infected heads can be made in wheat and durum fields to estimate the percent infection in the crop. Counts should be made at flowering time, as loose smut heads are hard to detect later. The counts provide information on the percent infection in the seed that was planted but are not a reliable estimate of the amount of infection in the seed of the harvested crop.

If weather conditions are favorable, *i.e.* cool and wet, infection in the harvested seed could be considerably higher than in the seed that was planted. Carboxin, difenoconazole, tebuconazole, and triadimenol seed treatments control wheat loose smut.

Scab and Black-point

Scab and black-point are fungal diseases that attack both wheat and barley seed. The scab fungus also may infect oats. The scab fungus infects the kernels during flowering if warm, wet conditions prevail. Black-point infection occurs from heading to maturity and also is favored by warm, wet weather.

Scabby seed is shriveled, light in test weight, and often has a chalky white or pink discolouration. When planted, scabby seed has poor germination and poor vigor. The scab fungus does not grow systemically from the seed through the plant to cause subsequent head scab infection. The source of head infection is spores from infected small grain or corn residue.

Black-pointed seed has a black to brown discolouration of the embryo or germ end, a discolouration which may extend around the kernel into the crease. When planted, black-pointed seed also may have low germination; if the seed germinates, the seedling roots are often infected by the fungus.

Protectant seed treatments or protectants in combination with systemics, or systemic products alone have provided significant improvement in stand and vigor. Yield increases also have occurred with treatment of scabby and black-pointed grain. Seed treatment helps assure good stands and seedling vigor; it will not prevent possible scab or black-point infections of the head during the growing season.

Take-all

Take-all is a root disease of barley and wheat caused by a fungus that thrives in very wet soils. Take-all generally is not a serious or common disease in barley or wheat in North Dakota, unless the crop is grown under irrigation. Triadimenol (Baytan), difenoconazole (Dividend) and tebuconazole (Raxil) are registered for suppression of this disease.

PYTHIUM

Damage to wheat from the *Pythium* fungus has not been well documented in North Dakota, but the fungus is common in agricultural soils and generally does more damage where wheat is planted into crop residues and soil temperatures are cool and soil moisture is high. Several seed treatment fungicides are registered for control of *Pythium* spp. in wheat, including difenoconazole + mefenoxam (Dividend XL), mefenoxam (Apron XL), and metalaxyl (Allegiance).

OATS

Smuts on oats have not been reported as a recent problem in North Dakota, although they have been serious in other states in the region. Oat smut can cause severe losses. Distinguishing between covered and loose smuts of oats in the field is not easy.

Oat loose smut is more difficult to control with protectant fungicides than oat covered smut, because the loose smut spores are lodged under the hulls where they are difficult to reach. Carboxin and tebuconazole are registered for oat loose smut control.

LEGUMES

Alfalfa and Small Seeded Legumes

Alfalfa, clover and other small seeded legumes may be treated for seedling blight diseases and damping off. The extent of occurrence of these disease problems in North Dakota has not been documented.

Captan or thiram products are registered for seedling blight control, while mefonaxam, metalaxyl, and oxadixyl products are registered for control of *Pythium* damping off and early season infections by *Phythophthora.*

Soybeans

Generally, seed treatment is not required for soybeans. However, it may pay to treat seed if: the germination is below 85 percent; the seeds are contaminated with fungi; the seeds are badly weathered or injured; or the seed coats are broken. If captan or PCNB treated seeds are to be inoculated with *Rhizobium* bacteria, the inoculant should be applied in-furrow.

Thiram, fludioxonil, metalaxyl and mefenoxam have little or no effect on *Rhizobium* bacteria. Generally, inoculant is not used on soybeans planted on land that was previously cropped to soybeans.

In fields with *Rhizoctonia* problems, a seed treatment containing carboxin, chloroneb, or PCNB should be used. Seed infested with Sclerotinia should be treated with fludioxonil, which will control infection of seedlings by the Sclerotinia fungus.

Dry Edible Beans

Most dry bean seed is treated prior to sale, often with a combination of fungicide, insecticide and bactericide. A fungicide protects against seed- and soil-borne fungi. The bactericide streptomycin controls surface-borne blight bacteria but will not control internally-borne blight bacteria.

Chickpeas

Several seed treatment fungicides are registered for use on chickpeas. Fludioxonil (Maxim) gives broad spectrum protection against soil-borne fungi; mefenoxam and metalaxyl (Allegiance) provide very specific and highly effective protection against *Pythium* seed rot and seedling blight and early season *Phytophthora* root rot.

Chickpea growers in some parts of western North Dakota have experienced severe difficulty with seed rot of untreated seed and should treat seed with mefanoxam or metalaxyl.

The most serious foliar disease of garbanzo beans is *Ascochyta* blight. *Ascochyta* can be seed-borne, usually at low levels. Wet weather may cause rapid spread of *Ascochyta* from a few infection centers. Thiabendazole (TBZ) is registered for use on garbanzo beans for eliminating seed-borne *Ascochyta*. Purchase of western grown seed that has been laboratory checked for *Ascochyta* also is desirable to minimize the danger of losses from this serious disease, but it should also be treated with thiabendazole.

Lentils

Lentils should be treated with captan, fludioxonil (Maxim), mefenoxam (Apron XL) or metalaxyl (Allegiance). Captan and fludioxonil provide broad spectrum protection against *Rhizoctonia* and *Fusarium* seedling blights. Mefenoxam and metalaxyl (Allegiance) provide excellent control against *Pythium* seedling blight, but do not provide protection against *Rhizoctonia* or *Fusarium*.

The two most serious diseases of lentils are anthrac-nose and *Ascochyta* blight. Both can be seed-borne at low levels but seed transmission to developing seedlings has not been deomonstrated for anthracnose. Ascochyta can spread rapidly in wet weather from a few infection centers.

The *Ascochyta* that infects lentils is a different species than the one that infects chickpeas. No fungicide currently registered is effective for elimination of either of these disease fungi on the seed. Purchase of western grown seed that has been laboratory tested for *Ascochyta* is desirable to help minimize the danger of losses.

Flax

Flax seed can be attacked by seed- and soil-borne pathogens, especially when the weather is unfavorable for germination and growth or when the seed coats are cracked and split. Seed treatment with protectant fungicides captan, mancozeb, maneb or thiram will reduce the amount of seed rot and seedling blight.

Yellow-seeded varieties are more susceptible to seed coat damage than are brown-seeded varieties. All varieties currently being grown are brown-seeded varieties except Omega, which is yellow-seeded.

Peas

Common seedling blights of peas can be controlled with captan, fludioxonil, PCNB or thiram. The water mold fungi *Pythium* and *Phytopthora* are common problems on peas and can be controlled with mefenoxam, metalaxyl or oxydixyl treatments. No seed treatment is registered for Aschochyta blight on peas, but data from Manitoba indicates that thiram provides good suppression of seed-borne Ascochyta.

Safflower

Safflower rust is borne on the surface of the seed and produces infections on the hypocotyl of the emerging seedlings. Seed-borne rust can result in poor stands and reduced vigor of the seedlings. Fungicidal seed treatment of safflower with one of several available fungicides is recommended to control seed-borne safflower rust. Carboxin, mancozeb, and thiram are labeled for seed treatment use on safflower.

The winter spores of safflower rust can survive from one season to the next in the soil but will not survive to the second season. Control of safflower rust requires seed treatment with a fungicide and crop rotation. Never plant safflower on land that had safflower the year before.

Sugarbeets

Pythium, Aphanomyces, and *Rhizoctonia* are fungi that may cause stand establishment and seedling disease problems in sugarbeets. The *Pythium* fungus occurs in most sugarbeet soils. It causes seed rot, pre-emergence damping off, and post-emergence damping off.

Post-emergence damping off caused by *Pythium* may occur when the seedlings are so tiny that they dry up and blow away within a day or two. Consequently, Pythium-induced seedling death is seldom noticed by the grower. The only thing noticeable may be an unusually poor emergence and stand.

Aphanomyces and *Rhizoctonia* cause death of seedlings at a later growth stage, with plants progressively dying from the two to the eight leaf stage. Later in the season, both fungi may cause a root rot which weakens the plant and also reduces the weight and quality of beet roots. Both diseases are favored by warm soil conditions. *Aphanomyces* is favored by heavy soils with poor drainage, resulting in saturated or puddled soils. *Rhizoctonia* is favored by moist soils.

Most sugarbeet seed is sold treated, but different treatments vary in their effectiveness against these three fungi. Certain fungicides have specific activity: mefenoxam, metalaxyl (Allegiance) and oxadixyl (Anchor) are highly effective against *Pythium;* thiram is moderately effective against *Pythium;* PCNB, chloroneb and fludioxanil (Maxim) are effective against *Rhizoctonia.*

Growers planting in fields with known *Rhizoctonia* problems may wish to request a special or supplemental seed treatment from their seed supplier or else use a planter box overtreatment. Hymexazol (Tachigaren) is effective against *Aphanomyces* and *Pythium.*

Canola

Blackleg is a fungus disease of canola that can cause severe losses. It is seed-borne and also spread by wind-borne as well as rain-splashed spores. The wind-borne spores come from canola crop refuse. Most long distance spread of the blackleg fungus is on seed. Most spread within a field or between fields is from wind-borne spores. The severe (highly virulent) strain of blackleg is common.

Benomyl seed treatment provides excellent control of the seed-borne phase of blackleg. Registration of other seed treatments effective against blackleg may occur in 2000 or 2001. Seed treatment is highly recommended for all canola seed planted in areas that do not yet have the severe strain of blackleg. In areas where the severe strain is prevalent, it may be necessary to plant tolerant varieties of canola. Other seedling blights may be controlled with captan, fludioxanil or thiram.

Sunflower

Downy mildew is a soil-borne disease that can cause severe losses if excessive rains occur shortly after planting. The downy mildew fungus is wide-spread across North Dakota. It survives many years in the soil and infects emerging or recently emerged seedlings when the soil is saturated. Several new races have occurred in recent years and at present only one hybrid is resistant to all races.

The downy mildew fungus has developed resistance to mefenoxam, metalaxyl and oxadixyl. Seed treatment with these products provides poor to fair control of downy mildew, depending on what percent of the downy mildew population is resistant. No suitable replacement fungicide was available for the 2000 growing season.

BIOLOGICAL CONTROL

Kodiak

Kodiak concentrate contains *Bacillus subtilis* bacteria which colonize the developing root system, suppressing disease organisms such as *Fusarium, Rhizoctonia, Alternaria* and *Aspergillus* that attack root systems. When used with a chemical seed treatment, the combination of chemicals and Kodiak provides protection to the root for a much longer time than with chemicals alone.

As the root system develops, the bacteria grow with the roots extending the protection throughout the growing season. As a result of this biological protection, a vigorous root system is established by the plant, which often results in more uniform stands and greater yields. Registered for seed and pod vegtables, soybeans, wheat and barley, and corn plus all other agricultural seeds.

SEED QUALITY

Quick seedling emergence and even stands are essential to maximizing the yield of all crops. The use of high-quality, disease-free seed is the first step to producing good stands.

The importance of using high-quality seed continues to increase due to changing production methods such as early planting, narrow rows, and no-till or reduced-tillage planting systems. The use of high-quality seed better ensures seedlings which emerge rapidly, tolerate adverse weather conditions, and resist disease.

SEED QUALITY AND SEED TREATMENTS

A number of factors affect the quality of seed. Seed-borne fungi, insect feeding damage, poor seed storage (temperature and humidity too high), and damaged seed coats can reduce the quality of seed.

Seed quality may be reduced due to infection by fungi when harvest of the seed crop is delayed several weeks or when seed diseases occur due to wet conditions. Infected seed can be detected at harvest as discoloured, shriveled, or moldy in appearance.

Insect feeding that may occur in the field prior to harvest or in storage may affect seed quality. Germination of seed lots may be reduced by the insect partially eating the seed or opening the seed coat to fungal infections occurring at a later date.

The seed coat is the primary protective layer surrounding the seed. Seed coat damage can result from rough handling during harvest, and cleaning or planting operations. Small cracks in the seed coat increase the chances of seed-rot by permitting water-soluble nutrients to escape into the soil which activate soilborne fungi that quickly enter the damaged seed.

Grain intended for seed should be handled as little as possible, dried using low temperature drying, and stored at moisture levels below 13 percent. High humidity and temperature during storage promotes growth of fungi and insect damage. Seed lots should be closely monitored for mold growth and insect activity during storage.

Seed treatments can play an important role in achieving uniform seedling emergence under certain conditions. The selective use of seed treatments can protect seeds or seedlings from early season disease and insect pests affecting crop emergence and growth. A few seed treatments may also be used to enhance crop performance during the growing season. Treatments containing Rhizobium inoculant have commonly been used to enhance the nitrogen fixing capability of legume crops.

Increased germination can be observed when seed treatments containing fungicides are used when poor germination results from a fungal infection. Fungicide are available to protect seed and seedlings from many seed-borne or soil-borne pathogens. Seed treatments are most beneficial for seeds infected prior to planting or when cool, wet soil conditions exist at planting which results in delayed germination.

Seed treatments containing insecticides can serve to protect seed and seedlings from certain insect pests found in the soil. Cool, wet, seed beds resulting in delayed germination or plowing down green manure crops may warrant including an insecticide seed treatment.

Seed treatments should not be considered a cure-all for the selection of poor seed lots. Seed treatments will not increase poor germination due to excessive mechanical damage, poor storage conditions, genetic differences in variety, or other damage.

The effect of a seed treatment on plant stands will depend upon seed quality as well as field conditions at planting. Extensive research has shown seed treatments can increase stands from poor quality seed or when planting conditions are less than optimum. However, seed treatments do not always increase yield. Generally, seed treatments result in increased yield by preventing death of seedlings causing frequent skips within crop rows.

Grain producers should realise that any one seed treatment will not control all diseases or insects that may attack seeds and seedlings. Protection is also short lived, generally only lasting as long as it takes for the crop to emerge.

Before selecting a seed treatment it is essential for the grower to consider potential problems associated with the seed and the history of problems associated with each field. Each planting situation is different requiring consideration of seed quality; planting rate; tillage and seed bed preparation; soil moisture and temperature; time of planting; and the likelihood of rapid emergence.

GERMINATION TESTING AND SEED QUALITY

The purpose of laboratory testing of seed germination is to assess seed quality or viability and to predict performance of the seed and seedling in the field. Seed processed for sale must be tested by a qualified laboratory under the Association of Official Seed Analysts Rules for testing seeds. Several different kinds of testing are available depending on the type of seed to be tested, the conditions of the test, and the potential uses of the seed.

The most common tests are the warm germination test, cold germination test, accelerated aging test, and the tetrazolium test. Each test is designed to evaluate various qualities of the seed. Factors that can affect the performance of seed in germination tests include; diseased seed, old seed, mechanically damaged seed, seed stored under high moisture, and excessive heating of seed during storage or drying.

In most cases a seed treatment will improve germination of seed only if the poor quality is due to seed-borne disease.

The most common test is a warm germination test because it is required by seed laws to appear on the label. The percentage of germinating seed in a warm germination test must be printed on the label of the seed if it is to be sold as seed. Germination is defined as: "the emergence and development from the seed embryo of those essential structures which are indicative of the ability to produce a normal plant under favorable conditions."

The warm germination test reflects the stand producing potential of a seed lot under ideal planting conditions. Usually 400 seed from each seed lot are placed under moist conditions on blotters, rolled towels, or sand and maintained at 77 degrees F for about seven days. At the end of this period the seedlings are categorized as normal, abnormal, or diseased, and dead or hard seeds. The percentage germination is calculated from the number of normal seedlings from the total number of seeds evaluated.

The cold germination test is designed to measure the ability of seeds to germinate under high soil-moisture content and low soil temperature. This vigor test simulates early season adverse field conditions and usually represents the lowest germination that would be expected from a seed lot planted under such conditions.

Actual field germination would normally fall between the cold test result and that of the warm germination test. Seeds are planted in a sand-soil mix at high moisture content and maintained at 50 degrees F for seven days. The test is then placed at 77 degrees F for four days. The percentage of healthy seedlings that emerge at least one inch above the soil is reported.

The accelerated aging test estimates the carryover potential of a seed lot in warehouse storage. The seeds are exposed to high temperatures and high relative humidity for short periods of time that cause seed deterioration. Seed are suspended over water in a chamber for 72 hours (wheat and soybeans) or 96 hours (corn) then tested in a standard warm germination test. This test only would be used on seed whose longevity was in question.

The tetrazolium test is a "quick test" for seed lot viability. It is useful when an approximate germination percentage is needed immediately. Seeds are soaked overnight in water then treated with tetrazolium to give an indication of viable, abnormal, and dead seeds in the seed lot. This test will not detect seed-borne disease, thus is limited in its ability to estimate seed quality. This test is highly reliable for determining viable seed of corn, wheat, oats, barley, and other grasses.

If growers wish to use bin-run seed and has not had a germination test conducted by a competent laboratory, they can get an approximate germination test using the following procedure in their own home.

Place two paper towels in the bottom of a dish or tray, one on top of the other; wet the towels thoroughly and tilt the tray up on one end so that excess water runs off the tray. Select a random sample of 100 seeds from the seed lot and place them in between the moist paper towels. Put the tray in a plastic bag and tie the end shut to prevent the towels from drying out.

Place the tray in a location of diffuse, not direct light, such as a north window. The location should be warm enough for good plant growth. An ideal location is with well growing house plants. After five to seven days, open the plastic bag and count the number of germinated seed with intact tap roots and shoots. Do not count moldy seed or diseased seedlings. Testing 400 seeds in this way will give a good indication of the germination percentage.

Choose the Right Fungicide for the Problem

The basic component of most seed-treatment products is a fungicide. Insecticides are contained in some products to give insect protection when such control is desirable. The grower should select a seed treatment product that will control the most likely pest to be encountered within a given field. It is important to note that not all fungi are alike. One fungicide seed treatment may be very effective against one fungus but will have no effect on another. Rates of active ingredient used also affect a seed treatment performance on different disease problems.

Fungicides have fungicidal activity, that actually kills the fungus, or fungistatic activity, that slows the growth of the fungus to prevent extensive damage. Fungicides used in seed treatment are of three types: protectants that generally prevent infection of seeds and seedlings by soilborne fungi, surface disinfestants that kill fungi and fungal spores on the surface of the seed coat (seed-borne fungi) and "systemic" disinfectants that kill fungi already established within the seed (seed-borne fungi).

Insecticides used in seed treatments act as surface protectants killing insects as they feed on the seed or emerging seedling.

The marketing of seed treatments by manufacturers may result in the same chemical being sold under several brand names and by several different companies. All chemicals have three names on the packaging; chemical name, common name, and trade or brand name. The chemical name represents the molecular structure of the active ingredient and is often long and complex. The common name is used by the manufacturer in lieu of using the longer chemical name. The trade or brand name is used by the manufacturer in product marketing.

The chemical name (R,S)-2-[(2,6-dimethlyphenly)-methoxyacetylamino]-proprionic acid methyl ester has the common name mefenoxam and is marketed under the trade or brand name Apron XL LS.

Using mixtures of different seed treatment fungicides or insecticides is a common practice. Mixing two or more fungicides or insecticides results in a product that will control a broader spectrum of seed pests.

Applying Seed Treatments

Seed treatments come in a variety of formulations: dry flowables (DF), liquid flowables (LF), true liquids (TL), emulsifiable concentrates (EC), dusts (D), wettable powders (WP) and others. These materials may be used in slurry and mist type seed treaters upon mixing with water as specified in instructions on the product label.

Special adhesives are incorporated into the dust formulations to adhere pesticide particles to the seed surface.

Commercial Treatment

Commercial seed treatment is often desirable due to the specialized equipment required to properly apply treatments or to treat large volumes of seed. The primary concern of the commercial treater is equipment calibration to ensure delivering the proper amount of active ingredient on the seed. This has become especially important with more modern fungicides that require only very small amounts of material (less than 0.5 oz) per hundred weight of seed.

Detailed information on the process of treater equipment calibration can be obtained from the equipment manufacturer. Seed treatment manufacturer labels provide information related to mixing formulations and mixing procedures.

Information needed to properly calibrate a seed treater includes:

1. The correct labeled rate of the seed treatment material that will be used,
2. How many pounds of seed is dumped each time the weigh pan arm dumps,
3. The size of the cup being used in the treater.

This information will allow you to calculate the correct application rate for the seed and treatment material used.

A new generation of seed treater is becoming available that offers precision distribution of treatment chemicals on individual seed through atomization. These advanced controlled droplet application (CDA) machines offer complete seed coverage with lower water volume, shorter drying time, and continuous seed flow.

These machines also reduce chemical exposure to employees and are capable of applying current and future products that may include both chemicals and biologicals.

On-farm Application

Many seed treatment materials are available for on-farm use. These are known as hopper-box or planter-box treatments. Most hopper-box treatments are dry treatments formulated with talc or graphite which adheres the treatment chemical to the seed. Liquid hopper-box treatments are available as a fast-drying formulation. Good seed coverage is required for maximum benefit from any seed treatment formulations.

Obtaining thorough seed coverage can be difficult when attempting to treat seed in the field at planting time. Most seed treatment companies recommend that the planter box be filled only half full with seed, and half the required amount of treatment product added. Mix the product into the seed volume with a stick. Then add the remainder of seed to the planter box and add the rest of the required amount of product, again mixing thoroughly. Good results are also obtained by premixing seed with the product in a suitable container, then pouring treated seed into the planter box.

On-the-farm treaters are available that apply liquid or dry formulations to seed as it passes through an auger from the transport bin or truck to the planter boxes. These treaters are a very convenient way to apply seed treatment onto bulk seed. For additional information on on-the-farm treaters, contact a representative of the company selling planter-box treatments.

Labeling Treated Seed

Once seed is treated it can no longer be used for food feed or other purposes. Under the federal and state seed laws, the following information is required to be shown on the label: a word or statement in type no smaller than eight points indicating that the seeds have been treated; the commonly accepted, chemical (generic) name of the applied substance including the application rate, and a caution statement if the substance used in such a treatment has an amount remaining with the seed that is harmful to humans or other vertebrate animals.

Seed treated with highly-toxic substances shall be labeled in red ink to show a statement such as "Poison Treated." In addition, the label should show a representation of skull and crossbones. Seed treated with less toxic substances shall be labeled to show a caution statement in no smaller than eight point type such as "Do not use for food, feed, or oil."

Colouring Seed Treated with Poisonous Substances

All treated seed must be coloured an unnatural colour to distinguish it from untreated grain and prevent inadvertent use as food for man, feed for animals, or for oil purposes. All treated seed must be coloured with a dye, colour coat, or colour film which must cause no apparent injury to seed germination or danger to personnel processing or using the seed.

Dyes are designed to stain the seed with colour. Colour coat pigments cover the seed with colour, and colour films are made from a polymer that creates a coloured film around the seed. Seed treatment companies currently offer a wide variety of colour and surface texture options to seed processors.

SAFETY PRECAUTIONS

Seed treatment chemicals are pesticides and should be handled as such. Product labels provide information on safe handling and application. Be sure to follow all instructions as listed.

Hazards associated with Handling, Sorting, Mixing and Storage of Treated Seed:

Read the label and follow instructions carefully. Over treatment may injure the seed and under treatment will not control the pests in question. The label will also provide information on the toxicity of the active ingredients in the form of code words.

The label will provide basic information on first aid, environmental hazards, directions for use, proper storage and disposal of containers, and pesticide and general product information. Use the label and be familiar with it. Material safety data sheets (MSDS) are available from the company for each of the products sold. Consult the MSDS for more detailed information before applying the material.

When handling toxic substances, operating personnel must be provided with protective clothing such as coveralls, cap, protective glasses, rubber apron, rubber boots, rubber gloves, and a respirator designed for use with the substance. Personnel should not inhale the dust or vapour nor permit the material to contact the skin or eyes.

The operator should wash thoroughly with soap and water before eating and smoking. Bathe immediately after work and change all clothing and wash clothing thoroughly with soap and hot water before reuse. In case of contact, immediately remove contaminated clothing and wash skin thoroughly with soap and water. A safety shower should be installed in the immediate vicinity of the treater.

An exhaust system should be installed to remove vapours and dust from the operating area. The exhaust air should discharge into a cyclone or bag-type dust collector. The treater should also be vented and tied into the exhaust system. Seed treaters should be isolated and not operated in the vicinity of other personnel or farm commodities that are to be used for food, feed, or oil.

Special multiwall paper bags or tightly woven bags are recommended for seed that has been treated with toxic substances. Seed should be thoroughly dry before going into the bag, as excessive moisture can cause rapid deterioration of the seed. Store treated seed in a cool, dry place away from food or feed products.

Seed treaters should be thoroughly cleaned after use as some of the pesticides are corrosive and others settle out and cause clogging of the equipment. Do not run contaminated water into a stream or public sewer, but discharge into a shallow ground pit.

Hazards associated with the use of treated seed and the disposal of unused treated and Empty Pesticide and Treated Seed Containers:

The user of treated seed should read and follow the label carefully. The planter hoppers should be filled outdoors. Do not breath dust or fumes from treated seed or get in eyes or on skin. Wash thoroughly with soap and water after handling treated seed. Use treated seed as seed and do not introduce them into the food or feed channels because serious injury to livestock and humans may occur. Follow instructions on the label for disposal of seed treatment containers and bags containing treated seed.

Do not reuse pesticide containers. Triple rinse and, if possible, recycle containers. Containers can be punctured and disposed of in a sanitary landfill or incineration, or, if allowed by state and local authorities, by burning. If burning, stay out of the smoke. Old seed with low germination should be destroyed by burying at least 18 inches deep in an isolated area away from the water supplies. Treated seed exposed on soil surfaces will be hazardous to birds and other wildlife.

Biological Seed Treatments

Recent research has focused on biological control of certain seed and seedling diseases of crops. The most promising biological control agents tested have been either fungi or bacteria. These biological seed treatments control seed pests by parasitizing the pest organisms, competing for food on the root system, or producing toxic compounds that inhibit pathogen growth. Several biological seed treatments are currently on the market.

Insecticide Seed Treatments

A limited number of insecticides are labeled for use as seed treatments to protect seedlings of various field crops from soil inhabiting insects.

SOIL INSECT PESTS

Soil insect pests are generally more of a problem on corn and soybeans than on small grains and forages. Soil-applied insecticides recommended for control of rootworm larvae also may prevent early stand losses due to wireworms, grubs, seedcorn maggot, and seedcorn beetles. If a soil-applied insecticide is used at planting, then a seed applied insecticide may not be necessary.

Refer to the label of the soil-applied insecticide to determine which insects are controlled before purchasing a seed treatment insecticide. Stand losses to these soil insects is generally not significant enough to warrant routine use of soil-applied insecticides in first-year corn. The use of insecticide for first-year corn may change in the future due to a biotype of the western corn rootworm beetle.

Benefits from suppression of early-season soil pests may be significant in:

1. No-till or reduced-tillage conditions with poor weed control where a high risk exists for a given insect problem such as wireworms and grubs in corn following established sod,
2. Continuous corn where early-season pests may cause unacceptable stand losses.

Seedcorn Maggot

Seedcorn maggot or the bean seed maggot are small, yellowish-white, legless fly larvae found feeding on corn seeds. Extensive feeding by these maggots will cause a reduction in stand.

In general, seedcorn maggot problems are most likely to occur in situations where:

1. High organic matter or decaying vegetation attracts egg-laying adult flies.
2. Cool, damp soil conditions delay seed germination and prolong the period vulnerable to maggot attack.

No-till production systems have not been found to increase the likelihood of problems with seed maggots. Several products are formulated for application prior to planting seed on the farm as planter-box or hopper-box treatments.

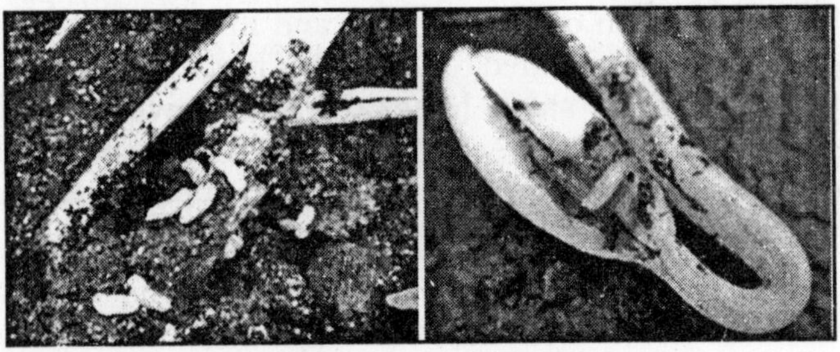

Fig. Seedcorn maggot

Seedcorn Beetle

Fig. Seedcorn beetle

Partially eaten seeds, loss of germination, or stunted seedlings in the presence of small (1/4 to 1/3 inch long) brown ground beetles indicate a seedcorn beetle problem. As with seedcorn maggots, damage is most likely to occur under cool, damp conditions where seed germination and seedling development are delayed.

Wireworms

Wireworms are the larval stage of the "click beetles." The term "wireworm" applies to a complex of species with life cycles requiring one or more years per generation. Wireworm populations affecting corn are most severe in fields following sod or fields having a prolonged grassy weed problem. The larvae pass through a number of life stages, or instars. The earliest stages are very small and white; the latter stages have a characteristic hard-shell appearance and are yellow-brown in colour.

Full-grown larvae range from 1/2 to 1 inch long depending on the species. Wireworm's damage corn by feeding on germinating seeds and young seedlings and may bore into stalks at the soil level. Stand loss may be significant in fields with high populations. In fields where a soil-applied insecticide is not used, especially first year corn fields, application of seed treatment insecticide is strongly recommended.

WHEN TO TREAT SEED

Corn

Treatment is recommended for all seed corn to prevent or reduce seed decay and seedling blights. The germination and early growth of the seedling is a critical period in the life cycle of corn. Soilborne organisms may invade and kill the embryo before germination, or when invasion occurs later, the seedlings may be destroyed before or after emergence. Seedlings surviving fungal infections at germination are often less vigorous.

Corn seedling diseases are more prevalent in cold, wet soil than when soil temperatures are above 55 degrees F. Therefore, early planted corn needs the added protection of a good seed treatment fungicide. Seedling blights tend to be more severe in fields with poor seed bed preparation such as in no-tillage or reduced-tillage fields. Corn seed is generally treated with a fungicide or fungicide-insecticide combination by the seed producer or seed processor.

Corn smut, leaf blights, stalk and ear rots, and virus diseases are not controlled by seed treatment fungicides.

Small Grains

Seed treatment of small grains is recommended to control smut diseases and to reduce seed decay and seedling blights. The smut diseases include stinking smut (common bunt) of wheat, loose smut of wheat and spelt, and the loose and covered smuts of oats and barley.

All of the wheat varieties planted are susceptible to stinking smut (common bunt). Wheat and oat varieties have resistance to some, but not all, races of loose smut (wheat) and loose and covered smuts (oats). The only sure method of smut control is by using an effective seed treatment fungicide. Spelt is extremely susceptible to loose smut and all seed should be treated with an effective fungicide.

Seedling blight phases of head scab (*Gibberella zeae*) and glume blotch (*Stagonospora nodorum*) contribute to poor stands of small grains. Both seed infecting fungi contribute to poor quality seed by causing lightweight, shriveled grain. Since these seed-borne diseases are common, all small grain seeds should be treated to control seedling blight when infected seed are planted.

Soybean

Seed treatment for soybean is recommended in three disease situations:

1. When poor quality seed is used for planting,
2. When early season control of Phytophthora damping-off is needed,
3. When planting early into cool, wet soils, especially in reduced tillage.

The major cause of poor-quality soybean seed is Phomopsis seed rot. Infected seed may be visibly moldy, yet others appear healthy. Purchasing certified seed with the percent germination, the variety name, and the seed treatment material used listed on the label is one way to ensure quality seed is planted. Germination percentage of bin-run beans is generally unknown unless they are tested.

Seed lots with greater than 80 percent germination generally do not benefit from seed treatment when planted at recommended seeding rates in warm soil. Those with less than 80 percent germination may benefit from seed treatments by increased stand and, possibly, increased yield. Decisions to use poor-quality seed should be based on the fact that any seed treatment fungicide effective against seed-borne diseases will not increase germination more than 20 percent.

If bin-run beans of 50 percent germination are treated with a fungicide, the grower should not expect over 70 percent germination in the field. Planting rates can be increased to compensate for seed of low germination.

A number of fungicides are effective in controlling Phomopsis seed rot. These materials have varying degrees of activity against seed and seedling diseases.

Phytophthora damping-off is a serious disease of soybean seedlings in the more heavy, poorly-drained soils. The increase in no-tillage and reduced tillage has increased the incidence of Phytophthora damping off in the state. No-till soils remain wetter longer in the spring and less precipitation is required to saturate them compared with plowed soil during the early stages of crop development.

Extra moisture in the soil pore spaces favors germination of *Phytophthora*. It is recommended that seed treatments be used even when planting *Phytophthora* resistant varieties to ensure good stand establishment.

Rhizoctonia seedling blight has caused considerable stand losses over the last decade. Damage from *Rhizoctonia* appears to be a stress related in that factors that adversely affect germination and establishment of seedlings predisposes the plants to infection. Stress from planting in dry soil, herbicide injury reduced tillage fields and shallow planting appears to increase the potential for damage by *Rhizoctonia*. Adequate seed bed preparation, soil fertility, and soil moisture for rapid emergence lessens the potential for problems from *Rhizoctonia*.

Sclerotinia stem rot may be introduced into new fields by sclerotia-contaminated seed lots or infested seed. Seed should be well cleaned to remove sclerotia and soybean seed should be treated with a fungicide to eradicate the fungus on infested seed.

Alfalfa and other Small-seeded Forage Legumes

Although a number of fungicides are registered for use as seed treatments of alfalfa, clover, etc., these crops usually do not respond to seed treatments under field conditions. Some increases in stand may result but these increases have not been reflected in increased yield. The one notable exception to this is the use of fungicide seed treatment on alfalfa for control of seedling damping-off caused by *Phytophthora*.

Significant improvement in stand establishment of alfalfa has been noted in research plots using treated seed. In areas where Phytophthora root rot has caused a serious problem in reducing alfalfa stands or preventing establishment of alfalfa, a variety with resistance to *Phytophthora* that has been treated with a *Phytophthora* specific seed treatment is highly recommended.

Rhizobium Inoculation and Seed Treatments

Rhizobium and Bradyrhizobium are bacteria which can fix nitrogen from the air or collect nitrogen from the soil water solution and incorporate it into compounds that plants can use. The nodules that form on soybean roots are the sites where the bacteria reside. Rhizobia are present in soils, but the commercial strains and some newer strains identified by the USDA are much more efficient at fixing nitrogen. The bacteria may be applied to the seed or placed in the seed furrow at planting.

Insecticide and fungicide seed treatments can have a negative impact on inoculants applied to soybean seed. Several factors affect the impact of seed treatments on seed applied inoculants including: toxicity of material, toxicity of carrier or formulation, length of time inoculant is in contact with seed treatment, and formulation of inoculant. Insecticide seed treatments tend to be the most toxic of all seed treatments to inoculants. The fungicides themselves are not necessarily the problem, but the formulation or carriers may inhibit Rhizobial growth and colonization.

Using fungicides in combination with inoculants can be successful if some precautions are taken. If a specific material is highly toxic to inoculants, then placing the inoculant directly in the furrow and not on the seed is recommended. Treating the seed first and allowing the seed treatment material to dry followed by using humus preparation of inoculant may be successful.

Liquid inoculants and fungicides should not be mixed and applied simultaneously. Producers should minimize the time the inoculants are in contact with the seed treatment fungicides prior to planting.

Corn Diseases

Damping-off and seed decay. Germinating corn seed can be attacked by several seed-borne and/or soilborne fungi. Pre-emergence and post-emergence damping-off caused by fungi is most common in poorly drained, cold soils. Seed rots and seedling blights are commonly caused by *Pythium* and *Fusarium* species, and may be caused by *Penicillium* and *Bipolaris* species. All of these fungi can rot seed prior to germination. Infected seedlings typically show a marked softening of stem tissue at the soil line.

Kernels with surface cracks caused by mechanical harvesting are especially susceptible to seed rots caused by soilborne pathogens. Broad spectrum, protectant-type fungicides are highly recommended for both corn and sorghum, especially when early planting in cold, wet soils is attempted. Pythium seedling blight is most significant under these conditions.

Most seed corn companies treat their seed prior to bagging. However, planter-box formulations of these materials are available for use by the grower on untreated field, sweet, and popcorn seeds.

Soybean Diseases

Phytophthora damping-off and root rot is caused by the fungus *Phytophthora sojae,* and is particularly a problem in low, poorly drained, clay soils. However, the disease can be encountered on a variety of soil types if the soil remains wet for several days soon after planting.

Phytophthora can attack soybean plants at any stage of development and stands can be reduced by seed rotting and pre-emergence damping-off. Young plants can be killed soon after emergence. Brown, water-soaked stems and yellow, wilted leaves are the primary symptoms of post-emergence damping-off.

Phytophthora survives in the soil as thick-walled spore, called oospores. Early in the growing season, when the soil remains wet for several days, these spores will germinate, producing a second type of spore called a sporangium. The sporangium then germinates to produce a third type of spore called a zoospore in large numbers.

Zoospores are attracted to chemicals released by soybean roots and swim through a thin film of water until a root is encountered. The zoospores produce a hyphae (a threadlike structure) and infect the soybean root. The fungus grows in the roots and eventually into the plant stem. As the plant dies, oospores are formed by the fungus.

Pythium damping-off can be caused by several species of *Pythium*. Pre-emergence damping-off, seed decays, and root rots of soybean all can be caused by this group of fungi. Pythium rots can occur on soybeans at any stage of plant development but are much more prevalent on seedlings.

Diseases caused by *Pythium* can occur over a wide range of temperatures, but are most common during cold periods with high soil moisture.

Seedlings infected with *Pythium* often fail to emerge. Stems of infected seedlings appear water-soaked and translucent above the soil line. Diseased areas later turn brown and appear shrunken; eventually, the stem and smaller roots decay and the seedlings die.

Pythium species survive also as oospores. During periods of high soil moisture, the resting spore germinates and infects seeds or young plants in much the same way as *Phytophthora*. Germinating soybean seeds release a wide variety nutrients or by-products that can stimulate the growth and attract spores of the fungus. Many different seed treatment materials will effectively prevent losses from Pythium seed rot and seedling blights.

Rhizoctonia seedling blight is caused by the fungus *Rhizoctonia solani*, and may occur at any time, but a protected cool period of low soil moisture followed by warming soil temperatures with a brief rainy period favors disease development. Seedlings under stress are more susceptible to Rhizoctonia seedling blight. Example of stress are herbicide injury or desiccation during early seedling growth.

Seedlings and young plants are most often affected. Infected plant stems and older roots appear reddish brown and have sunken lesions. Stems may be girdled just above the soil line; tissue thus damaged may appear cracked or cankered. In dry, windy weather, severely infected plants wilt and die rapidly.

Rhizoctonia solani survives in the soil as sclerotia. The fungal mycelium body itself may also survive in soil or in association with old plant residue. Growth of the fungus in soil depends on soil nutrient supply, pH, moisture, and temperature.

As the fungus grows in soil, it can encounter and infect germinating soybean seeds or young plants. Several seed treatment materials can control the seed rot phase of this disease, however, none are highly effective against the post-emergence or stem rot phase of the disease.

Phomopsis seed rot is a common seed disease and is caused by the seed-borne fungi, *Phomopsis longicolla, Diaporthe phaseolorum* var. *sojae,* and *D. phaseolorum* var. *caulivora.* Infected seeds typically germinate poorly or not at all.

Fig. Phomopsis seed rot of soybean

.Severely infected seeds appear shriveled, cracked, and elongated, and may be covered with a white, moldy growth. However, seeds may be infected and fail to show any symptoms at all. Infection of seeds is most common when warm, wet weather delays harvest. When moldy seeds are planted, seedling death results in poor stands.

Generally, higher soil temperatures favors development of seed rot and seedling blight. Seed treatments can increase the germination rate up to 20 percent. Seed lots with less than 80 percent germination should not be used.

Sclerotinia

Recent research demonstrates that Sclerotinia white mold can be introduced into new fields on infested soybean seed. In addition, this fungus, *Sclerotinia sclerotiorum,* forms hard, black, small, irregular-shaped sclerotia both inside and outside of infected soybean plants.

These sclerotia are harvested with the seed or returned to the soil with plant debris at harvest. Seed should be well cleaned to remove sclerotia and treated with appropriate fungicide seed treatment to eradicate the fungus from infested seed.

Wheat and Spelt Diseases

Common bunt or stinking smut is caused by the seed-borne fungus *Tilletia laevis*. The disease causes reduced wheat yields and grain quality. Common bunt was once the most prevalent disease of wheat, however, today the disease is rarely found due to the widespread use of effective seed-treatment fungicides.

Heads of diseased plants may appear stunted and are noticeably thinner and darker in colour than healthy, disease-free heads. "Bunt balls" (kernels containing millions of fungal spores) replace the kernels in the flowering heads and spread the glumes much farther apart than normal. Bunt balls approximate the size of normal kernels but tend to be more spherical and off-coloured (typically gray-coloured). When the bunt balls are ruptured at harvest, they release spores into the surrounding air and contaminate the surface of healthy seed.

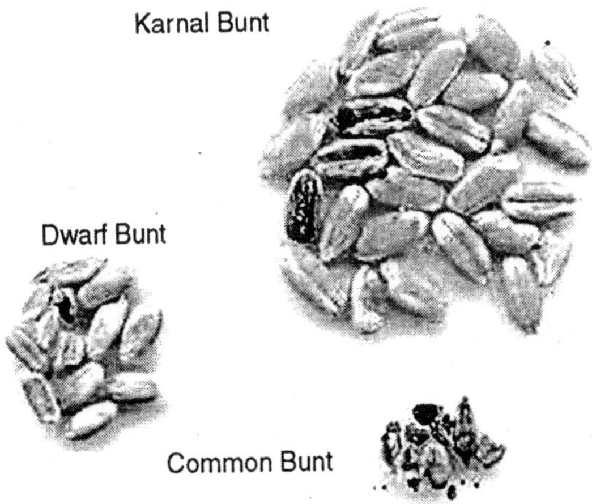

Fig. Bunt of wheat

Bunt fungi survive as resting spores on contaminated seed. When contaminated seeds are planted, bunt spores germinate in the presence of moisture and infect the wheat seedlings. Once established, the fungus grows systemically in the plant. At flowering, entire flowering heads may be infected, and the fungus converts the developing kernels into bunt balls. Healthy seeds are contaminated by the release of the spores at harvest.

Loose smut, caused by the fungus *Ustilago tritici*, has little effect on seed quality. However, it can result in substantial yield losses if left unmanaged. Symptoms are most noticeable between heading and maturity of the crop. Diseased flowering heads are conspicuously darker when compared to healthy, green heads. Also, diseased heads typically emerge earlier. Diseased spikelets may be entirely transformed into dark fungal spore masses.

As the head emerges, floral tissue is torn and the spores are released. Eventually only a spike is left where a normally healthy head should be. Spores of the fungus are dispersed by wind to nearby healthy flowers where they initiate new infections during wet weather. Unlike bunt, the loose smut fungus infects the developing kernels without causing noticeable damage.

Ustilago tritici survives in infected wheat seed. The fungus becomes active when the seed germinates and grows into the growing point of the developing wheat plant. Fungal growth closely follows the plants' growth. When the flowers form, the fungus again sporulates and starts the cycle again. Use a systemic fungicide seed treatment to eliminate this pathogen from infected seed.

Head scab, caused by *Fusarium* spp., is frequently a major cause of poor quality wheat seed. Infected seed appear whitish to reddish in colour and are nearly always shrunken and light in weight. Scabby kernels may be dead or the young seedling may be killed by seedling blight after emerging from the soil. Roots of plants killed by seedling blight appear light to reddish brown in colour and may be covered with mold.

If they survive, they generally lack vigor and may produce only a few weak tillers. Planting infected seed early when soil temperatures are above 60 degrees F generally increases loss from seedling blight. The best control of the seed-borne phase of scab is to first clean seed lots to remove all lightweight seed, thus increasing test weights; and second, treat with a fungicide that is effective in controlling seedling blights.

Fig. Head scab of wheat

Stagonospora glume blotch, caused by the fungus *Stagonospora nodorum* has increased in the US in recent years probably due to expanded use of nitrogen fertilizers, semidwarf wheat varieties, and reduced tillage. Symptoms can develop throughout the growing season on all above-ground plant parts. Initial symptoms are small chlorotic flecks usually on lower leaves or those in contact with soil. The glumes of the wheat heads become infected in late spring before flowering. Lesions generally begin at the glume tips and are dark brown in colour. During favorable weather, the fungus penetrates the glumes and infects the seed causing severe shriveling of the grain.

Stagonospora survives on seed, straw, and/or volunteer wheat. With the onset of moist weather, spores are produced and are spread to healthy wheat plants by splashing rain. Glume blotch is most common in relatively warm weather when the temperature is between 70 degrees F and 80 degrees F.

Barley Diseases

True loose smut of barley is caused by the fungus *Ustilago tritici*. The disease cycle is exactly the same as that of loose smut of wheat. Infection occurs when the flowers develop. The seeds become infected internally and must be treated with a systemic fungicide to eliminate the pathogen.

Semiloose smut is caused by the fungus, *Ustilago avenae*. The disease typically causes small losses due to the use of resistant cultivars and fungicide seed treatment. It is most common when soil temperatures are cool and the soil is dry. Distinguishing semi-loose smut from true loose smut in the field is extremely difficult. Diseased heads typically appear earlier than healthy heads. Millions of fungal spores may be formed in diseased heads.

Ustilago avenae survives as teliospores on the seed surface. At seed germination, the fungus becomes active and infects young seedlings prior to their emergence from soil. As the plants develop, the fungus grows within the growing point of the young plant. At the boot stage, the fungus transforms the kernels and glumes into dark spore masses very similar to those produced by the true loose smut fungus. Spores are readily dispersed by wind or the action of harvesting equipment. Healthy seeds are contaminated upon contact with the spores.

Covered smut caused by the fungus, *Ustilago hordei*, can result in significant yield losses if left untreated. It is a relatively common disease throughout the barley-growing regions of the world. Problems with the disease occur over a wide range of temperatures and are most common under moist soil conditions.

Ustilago hordei survives as thick-walled resting spores on barley seed. Typically, spore germination coincides with seed germination, it is at this time that infection occurs. Once established, the fungus invades the actively growing tissue and keeps pace with growth of the plant. At flowering, the fungus grows through the floral tissue and forms masses of spores in place of healthy seed. Each kernel becomes a spore filled "smut ball" that is enclosed by a thin membrane until plant maturity. At harvest, spores are liberated and may contaminate healthy seed.

Barley stripe, caused by the seed-borne fungus, *Drechslera graminea,* is most common during periods of rain or high humidity. Few seeds are produced on infected plants, and losses generally coincide with disease levels in a particular field. Properly treated seed is the primary means of controlling this disease. Symptoms are first noticeable on the first two or three leaves produced by the plant and on many leaves produced later.

A yellow stripe on the leaf sheath and basal portion of the leaf blade is common. Later, the stripe may extend the entire length of the leaf and the leaf may die. At heading, diseased plants are often stunted and appear a light tan colour in contrast to the larger, green healthy plants. Spikes may be deformed or may fail to emerge. At heading, spores are produced on diseased leaves and spread by wind to nearby heads. Grain produced by diseased plants is typically brown and shriveled.

Drechslera graminea survives entirely as mycelium in the outer layers of the seed. When the seed germinates in moderately moist soil at temperatures below 55 degrees F, the fungus mycelium penetrates and infects the seedling. The fungus grows within plant tissue and occupies the actively growing plant parts.

Net blotch and spot blotch, common diseases of barley, are caused by the fungi, *Drechslera teres* and *Bipolaris sorokineana,* respectively. Net blotch is most common during rainy and humid weather, and early flag leaf infections can result in both reduced grain yield and weight.

A net-like pattern on sheath leaves or flag leaves is the most recognizable symptom of the disease. Initially, small spots or streaks appear on the leaves, followed by an expansion into conspicuous longitudinal and transverse streaks, eventually combining into a network. Brown areas surrounded by greenish-yellow margins are commonly observed on infected leaves. Eventually, entire leaves may wither and die.

Drechslera teres survives in infected seed or on crop residues. Infection of seedlings is most common at 34 degrees F to 60 degrees F. Once infection is established, the fungus produces spores on leaf tissue during periods of high relative humidity. Spores are released and carried by wind to nearby barley plants where new infections can result. Infection of leaves can occur when leaves are wet for periods of 5 to 15 hours within a 46 degrees F to 91 degrees F temperature range.

Spot blotch is most commonly a problem during warm, humid weather. Yield losses of up to 36 percent have been attributed to this disease when the flag leaf has become severely diseased early in the season and has died prematurely. Seedling blight (*B. sorokineana*) in winter barley is favored by planting early in warm soil. Delayed planting of winter barley in cooler soil helps prevent blight.

Small brown spots surrounded by a yellow halo is the characteristic symptom of the disease. Both leaves and leaf sheaths on plants of all ages can become infected. Eventually, the spots coalesce and cover large areas of the leaves. Minute fungal fruiting structures containing spores can sometimes be observed in the larger spots. Severely diseased leaves may eventually die.

Bipolaris sorokineana survives in seed or infested crop residue. Leaf infection results from airborne spores produced on the seed or on residue. Epidemics of spot blotch can occur when there are prolonged wetness periods (greater than 16 hours) and the temperature is above 70 degrees F.

Note that net blotch is more common in cooler, humid weather, as opposed to spot blotch which prefers warm, humid weather. Most common broad-spectrum seed treatment fungicides are effective in eliminating seed-borne *B. sorokineana* and *D. teres* as well as preventing losses from seedling blight caused by them.

Oat Diseases

Loose smut and covered smut, caused by *Ustilago avenae* and *U. segetum*, respectively. When disease outbreaks are severe, both yield and grain quality can be significantly reduced. However, fungicide seed treatments effectively control both diseases.

Symptoms of these diseases are very similar. Smutted oat panicles emerge from sheaths as olive-brown to black fungal spore masses. Smutted panicles remain more compact than healthy ones, and usually all spikelets on a diseased plant are affected. Smutted plants are often stunted and easily detected in a recently headed field of oats.

The thin membrane enclosing the spore masses of *U. avenae* usually rupture and disintegrate after panicle emergence as opposed to those of *U. segetum*, which can persist longer.

Alfalfa Diseases

Phytophthora damping-off and root rot is caused by the fungus, *Phytophthora megasperma f.sp. medicaginis*. Standing water, poor drainage, and generally wet soil conditions favor disease development. This disease can be lethal to seedlings and established plants.

Seedlings may fail to emerge or can be killed after emergence. A conspicuous yellowing of leaves, particularly lower ones, is a characteristic symptom. Infected seedlings often wilt. Seedling tap roots have dark brown to black lesions and may be rotted 2-4 inches below the crown. A yellow discolouration of internal root tissue is often present. Tap roots may appear discoloured.

The life cycle of *Phytophthora* on alfalfa is similar to that on soybean, however, the fungi are completely different and do not attack the other plant host.

Seed treatments are labeled for control of Phytophthora damping off. However, to protect stands beyond the seedling stage, alfalfa varieties with resistance to the fungus should be used.

Chapter 7

Seed Collection

In agriculture, collecting seed from superior parent stock has been practiced for thousands of years. This technique results in higher yields and environmentally durable plants. This concept, however, has yet to gain widespread acceptance in forestry practice. To maximize yields and quality of trees for plantations, agro forestry, and other uses, the following principles for collecting seed should be observed.

In forestry, as in agriculture, the quality of offspring plants improved populations will result if the seed used to produce them was collected from superior individuals, stands or orchards. Seed quality is measured in two ways. One, by the physical quality of the seed and secondly by the desired physical traits of the resultant mature tree. The benefits of using quality seeds, chosen from selected parent trees, are twofold: improved survival and greater economic returns.

Seeds from healthy, well-formed trees provide greater assurance that resulting stock will have good form, survive and better resist stressed conditions due to marginal sites, frequent cutting, or harsh climates. These adverse conditions typically weaken all but the strongest trees, making them vulnerable to insects, fungi, parasitic plants, and diseases.

More important, because of the long-term resource investment and land and labour commitment in forestry, high survival rates of good quality trees is a must. In agriculture, farmers can recoup their loses after a poor season, or even two. Trees, however, occupy a site for years. Therefore, any reduction in growth or quality from planting inferior stock represents a lost opportunity, which can be measured in time and capital, for as long as a tree occupies a particular site.

For the second benefit, economic return, the investment in selection will be more than compensated for by higher product yields over shorter rotations. Better yields result in more building materials, higher fruit production and quality, faster fuel wood growth, and quicker and more prolific regrowth after lopping.

Effective seed selection can create success, even though sophisticated techniques may not be used. Any attention given to seed source will be a measured improvement over current practices. In addition, careful seed handling practices must be included as part of the seed collection programme.

Such practices include transport, seed preparation, handling, cleaning, grading, and seed testing. Disregarding these practices will result in damaged seed or allow the inclusion of unwanted or excessive amounts of foreign materials, which may hinder germination and plant production.

PLANNING THE SEED COLLECTION STRATEGY

Too often in large planting programmes the task of seed collection is an afterthought, typically left until the last minute and done hurriedly by unskilled or untrained labour while final nursery preparations are being made. Small-scale planters usually have access to left-over seeds from large programmes. In either case, little consideration is given to seed quality, form, or location of the parent stock.

Sometimes, to satisfy the requisite seed volume, nursery manager's purchase bulked seed from villagers. However, villagers tend to collect from the nearest trees or stands, which generally include immature, diseased, distressed and otherwise inferior seed. And, if seeds need to be collected from the crown, villagers will choose the easy-to-climb trees, which are typically small and misshapen. Such trees would not normally be chosen as prime seed source candidates. Also, in some species a single tree will produce a large amount of seed in some years.

Such a harvest may satisfy the bulk requirement, but would be genetically dangerous because only one phenotype is represented. This technique causes problems for both large- and small-scale planters.

Approach, using untrained labour or collecting seed from a single tree, can be a critical mistake. Such practices often result in low survival, substandard trees, and a poor economic return for the time and work invested, nursery and site preparation costs. Instead of an impromptu search or purchase from unknown sources, any seed collection, both large-scale and small-scale, should be a developed programme of scheduled activities overseen by knowledgeable individuals.

This is especially true for those involved in large-scale planting programmes. Those involved in small-scale plantings can often request advice from nearby large-scale planters. These essential activities include selection of preferable parent trees; preparation of equipment including provisions for handling, transport, extraction, testing, grading, and storage; recruitment and training of workers and; seed sampling and quality monitoring.

A timeline and work plan for the collection itself must be developed in conjunction with overall nursery operations. More important of these is the selection of preferable parent stock. This can be accomplished by an able person who can distinguish between desired and inferior seed sources. This person, in turn, can supervise field collection crews during harvest. Incorporating many of these activities often requires the level of funding available to large-scale operations. Nevertheless, even the small-scale planter should apply the principles as best possible whenever collecting seed.

GENETICS PROVIDES THE BASIS FOR SEED SELECTION

Tree improvement programmes worldwide have shown that seed collected from trees with particular traits tends to produce trees with similar traits. One can correctly assume that if good parent trees are selected, then most of the resultant offspring will be of desirable quality. Yet despite this concept's proven value and wide understanding, it is seldom applied in the field. To correct this shortcoming, simple parent tree selection guidelines should be developed.

These would identify the desired traits and specify the parent trees that reflect those traits. Such traits might include straight trunks, fast growth, and better form. These guidelines are normally met in national or company research programmes, but following them will benefit all planters who are careful about the source of their seeds. The traits in turn are chosen to reflect the ultimate desired use of the tree, including building materials, windbreaks, shade, fuel wood, or erosion control.

At the same time, these guidelines should not be made cumbersome. For the benefit of those individuals collecting seeds, the guidelines could be a condensed set of simple criteria. Indeed, immediate benefits can be derived from using even the most basic concepts, such as selecting trees of good form and vigor and avoiding trees that are poorly developed, diseased, dying, or are isolated from others of the same species. These simple methods should be applied to large and small programmes.

While it is important to collect seed from the best parent trees it is also important to gather seed from several parent trees. Collecting seed from several parent trees assures diversity among seedlings being planted. Whether the programme is large or small, maintaining diversity is important and will help guard against pest epidemics. To obtain adequate diversity, seed lots should be collected from a number of trees within a species' range. Researchers have yet to determine an ideal number of trees per seed lot.

In plantations and along roads more emphasis should be placed on collecting from the best trees. In all cases care should be taken to collect seed from several trees representative of the best parent stock. Researchers have yet to determine an ideal number of trees per seed lot, i.e., those seeds that will be used in one planting. However, most agree that collection should not rely on only one or two individual trees. According to some sources, a minimum of 15 to 25 mother trees per bulked seed source is preferred.

Seed selected from a narrow sampling of trees should be avoided as this will limit genetic diversity. Keeping a proper distance between selected parent trees is also vital. During natural regeneration of some species, seeds fall near the parent stock. In time, inbreeding may occur, which will result in lower quality individuals. To reduce the chance of collecting a seed lot predominated by half-siblings, a 100-meter distance between collection trees of the same species is recommended.

This practice is especially critical to species that are lopped or coppiced as a method of harvest and regeneration. The temptation to forego a minimum number and spacing strategy is perhaps greatest during years of abundant seed production. Here, large seed quantities are available from relatively few trees. At such times, an even greater effort should be made to ensure genetic variety. Even better, abundant seed years can be opportune. When properly cleaned and prepared, seeds can be stored for several years. Stored volumes of good seed will help ensure supplies during leaner production cycles.

Where possible, all large-scale planting programmes should have at least two years' seed supply in storage. Conversely, leaner seed years present other challenges. During such years shortfalls will occur and the urge will be to collect for quantity, without regard to quality. If at all possible, seed collection during lean production years should be minimized.

Aside from genetic and site qualities, other considerations must be factored into the collection strategy. For instance, only mature seed from ripened fruits should be gathered. Harvest schedules must account for the different times at which fruits from various species ripen. Depending on species, maturation can last between two to six months. To help determine ripening, periodic surveys of the selected stands are suggested. Surveys should begin after flowering, as is the practice with cocoa. To prevent any false readings, specific sample trees should be marked and used throughout the process.

The surveys will also serve as an early indication of the season's expected seed volume and quality. In addition, field surveyors can detect early infestations of insects, disease, or other seed predators. With such information, the harvesting strategy and schedules can be adjusted to accommodate any foreseen hindrances.

Aside from desired end-use traits, the tree must adapt to its growing site. In particular, it must withstand periodic harsh conditions, such as drought, and be able to grow in poor soils. Physical traits alone may yield few clues as to site adaptability. Here, environmental distinctions may help.

The collection staff should note even the slightest variation throughout a species' range. Topography, soils, microclimate, associated vegetation, and man-caused factors such as perennial grazing may all influence how a species conforms to its surroundings. In overall location, a species typically grows best in the middle of its range and fares less well at the edges.

However, seed lots should reflect all site variations, thus assuring that no one habitat is relied on too heavily. But some cautions are needed when sampling for habitat. When associated plants are used as site indicator species, one must discern between natural conditions and man-made alterations. Continual grazing can heavily degrade a site, even to the point where it changes the vegetative makeup. In addition, extreme dry or wet areas should be excluded if not within the normal bounds of a species' requirements.

During collection of particular stands, efforts should concentrate on trees that comprise the stand's perimeter. These individuals produce better quality seed. Aside from natural areas, man-made stands should be reviewed carefully before selection as a seed source. Early reforestation efforts may have given little thought to seed collection, thus producing inferior parent trees to those found in natural stands. Or, the species planted on the site may not be well suited for the area.

Whenever possible, historical data or records should be referred to for such stands.

A Short list of Collection Tactics

- Collect seed lots from between 15 and 25 individuals that are spaced at least 100 meters distance from any other collection tree of the same species.
- Choose trees in vigorous health and avoid any that are diseased, suppressed, deformed, environmentally stressed, or in otherwise poor health.
- Collect from trees that are well formed and either dominant or co-dominant in the canopy.
- Avoid individuals that are isolated from others of the same species.
- Harvest only mature seed from ripened fruits.
- To ensure genetic variation, collect fruits equally from all parts of the crown—top, sides, and bottom—as these parts may have been pollinated at varying times from different sources.
- Collect throughout a species' normal habitat, noting variations in site.
- Man-made stands, including live fencing, plantations, or windbreaks, should be carefully reviewed as to their establishment before being selected as a seed source.

How to select Good Trees

Parent tree selection in the field is best done as a planned team effort coordinated by a trained person. This selection of preferred trees serves to eliminate errors that would otherwise be produced by last-minute collection efforts. In addition, a planned, timely programme allows for the full use of available people and resources. Development of the collection strategy includes a series routine screenings of potential parent stock. The series is broken into three progressive phases.

Phase one is done well before the onset of flowering. Given simple guidelines for site selection and appearance, general laborers search for appropriate stands. Locations of such are recorded, along with general descriptive information.

Phase two involves technical staff. Here, the identified stands are reviewed for their potential as a seed source. The review includes analysis of the site makeup including soils and topography, species composition, volume and spacing, and age, size and health of the trees. Initial selections are made as to potential parent stock. These trees are marked for further study in phase three.

In *phase three,* trained professionals examine the marked individuals for a final selection. The trees are analyzed for their physical characteristics, including their form, branching, growth rate, dominance, crown cover, health, and any clues as to their seed productive capabilities. Before the final candidates are chosen, additional criteria are again considered.

Preference will be given to those trees located in the stand's perimeter. Finally, selected trees are marked and their location noted. Monitoring will increase as flowering begins and continue through the seed harvest period. Monitoring includes volume and quality of the seed as it matures, in addition to any indications of pest infestations including insects and disease.

SEED HANDLING AND PREPARATION

Seeds are rarely taken directly from the collection site for immediate planting in the nursery.

Instead, they must first undergo a preparation that will ensure proper germination and help eliminate bad or damaged seed. Preparation includes all activities from collection to sowing. These activities cover transport, sorting, extracting, cleaning, testing, and grading. In addition, seed harvesting and sowing periods are often months apart, which requires adequate storage facilities.

The facilities must protect the seed stock from the weather and intrusive pests such as animals, insects or molds. In some instances, proper temperature control may also be required. This is especially true if seed is to be kept over a number of years.

Handling in the Field

Once the fruits are picked or collected, they must be transported directly for seed processing. Any lag time in the field could cause desiccation or predation by various seed pests. The seeds must also be well-protected from the weather to prevent excessive contact with moisture, which may prompt early germination. In addition, because of the variation in harvested fruits, handling techniques must flexible.

Those in charge of collection must have adequate knowledge as to each species' requirements and limitations during transport. While the hard coated seeds of most leguminous species travel well, other seeds are more fragile and require greater handling care.

Proper Conditioning Limits Seed Degradation

The nursery manager must have facilities readied for seed preparation at the time of collection. Such facilities should be located at the nursery site, thus eliminating the need for further transport before sowing. Seed extraction should occur as the fruits arrive from the field. Again, any lag between collection and preparation will cause the seed to degrade or spoil. Depending on species, equipment for both dry and moist extraction may be needed.

Dry extraction requires appropriate sorting and drying areas, tumblers, and screens for sieving. Moist extraction calls for a method of macerating the fleshy fruits, followed by drying and cleaning of any remaining foreign materials.

In either case, all inert materials must be removed as these can result in poor germination or hamper storage ability. Also, the rate of drying and moisture content must be carefully monitored during the drying process. Seeds dried too quickly may become damaged. While small-scale planters may not have access to sophisticated equipment for drying and cleaning seeds, they should aim to remove as much foreign material as possible.

After extraction, the pure seed is then graded. Grading helps ensure a more homogenous seed set, which in turn will produce more uniform growth in the nursery. Grading is done either by seed size, weight, or a combination of both. Actual seed size within the same species will vary due to a number of factors. The goal is to eliminate undersized, immature, or deformed seed. This usually can be done by hand if great quantities are not required.

However, graders must be provided samples of the desired well-graded product to use as a guide. In addition to grading, seeds are tested for germination rates, and are examined for phytosanitary concerns and species purity. Using these factors, the forester must develop criteria for consistent seed certification. No matter how carefully done, collection efforts in the field will be nullified if inconsistent methods of grading and testing are used from one seed lot to the next.

Proper Storage is Essential

If the prepared seed is not used within a reasonable amount of time, adequate storage must be provided. As noted, most leguminous seeds store well in normal conditions provided they are properly cleaned, dried, and protected. Such prepared, intact seed can be handled and stored without fear of damage. However, if storage is required for several years, special facilities, such as cold storage, may be needed for some species. On the other hand, if germination begins, different measures must be taken.

Here, the seed ceases dormancy and begins to use stored nutrients for survival. At this point, its shelf life is drastically reduced. Thus, it is important to closely observe the condition of the stored seed and monitor it periodically to ensure the storage environment remains intact.

Differences in how trees adapt to environments is often more important with a species than among species. There can be important differences in growth between seed sources from different geographical areas and environments. The areas and environments in which stands have developed through natural selection are call provenances. The growth, measured in height, in tropical hardwoods of the most vigorous provenances can be 30 to 50 percent greater than the growth in the least vigorous provenances.

Unfortunately, there are very few tree species that have been adequately explored in their center of origin and studied in provenance and progeny tests on multiple sites, leaving much uncertainty about provenances which are intended to represent an entire species. Perhaps the most striking example of this comes form the use of *Acacia mangium* in Sabah, Malaysia, where ultimately thousands of hectares were planted from seed collected from one tree. As a result, growth rates of third generation were one-half those of the first generation.

Seed Harvesting

Most wild seeds are collected by hand because the desired species usually do not grow in pure stands and the site's topography often limits the use of mechanical equipment. Plants are easiest to identify when they are flowering, so it is a good idea to mark individual plants with surveyor's flagging tape, or to write down specific landmarks to help you relocate populations when the seeds are ready. Always get permission from the landowner when collecting on private land. Never collect on public land. Areas destined to be developed or destroyed in the near future provide excellent sites for collecting seeds, provided the landowner has given permission.

Never collect seeds from rare or endangered species—collect only from plants that you find growing abundantly in a given area, to ensure that you do not eradicate an isolated population. Take at most, only one-tenth of the seeds so that enough seeds are left to reseed and perpetuate the stand.

Collecting seeds at the correct time is crucial for propagation to be successful. Gather fruits from the ground only if they have recently dropped. Reject fruits or seeds that have been on moist ground for some time, or any seed that may have begun to decay, mold, or become infested with insects. They could contaminate the rest of your seed harvest if combined with other seeds during storage. Delayed harvesting of species with persistent pods often results in insect-infested seeds.

The tools and material you will need depends on the size of the harvest. Basic equipment includes gloves, boots, drop cloths, pruning shears, boxes, baskets, paper bags, and/or canvas bags. Although plastic bags may be used for collecting, storing seeds in airtight containers or plastic bags will encourage mold growth. Many plants can be stripped by hand or the seed can be beaten onto drop cloths.

Seed Maturation

Proper seed harvesting is aided by an understanding of seed ripening, dispersal mechanisms, and the influences of weather on the timing of seed maturation. Flowering and fruiting dates vary from year to year. You should be familiar with the approximate flowering and fruiting dates and then be able to recognize mature fruit or seeds. An early spring and dry summer, may cause seeds to set early.

Seed quality also varies from year to year and from location to location. Experience is often the best teacher in learning to determine whether a seed is mature.

When seeds are mature you should begin collecting. Mature seeds are usually dark in colour, firm, and dry. Seeds that are green and moist are immature and generally will not germinate or will produce unhealthy seedlings.

The flesh of pulpy fruits often becomes soft and changes from green or yellowish to reddish or blue-purple when ripe. Seeds are often mature a week or more before the fleshy fruits turn colour and fall from the plant. You can determine seed maturity by cutting open the fruit and examining seeds for firmness, fullness, and dark colour.

A delay of only a few days may be the difference between success and failure in collecting a good crop, especially for those species with seeds that are dispersed quickly or are attractive to birds and other animals.

Many pods or capsules dehisce when ripe and mature at staggered intervals making a quantity difficult to collect. Once maturation begins on a plant, check it every few days to collect any newly matured seeds. You may also try inverting a paper sack over the immature seed and tying it off with string.

Seed Cleaning and Preparation

Seeds should be collected just before, or, as the pod or capsule turns brown and dries, and before it dehisces. The pods should be dried in single layers spread thinly on canvas cloths, screens, or trays elevated from the ground. Curing on the pod may take longer for species other than legumes.

Air-drying takes one to three days, depending on the humidity. After the seeds have dried, you can extract them from the pods by beating or thrashing. A mature pod will often twist and split open to drop the seeds.

Although not all seeds need to be cleaned before storage, those with pulpy fruit should be cleaned to reduce mold. Remove the pulp of large fruits by hand by rubbing on a screen or mashing with a wooden block, rolling pin, or fruit press being careful not to damage the seed.

You can clean smaller fruits with a blender, as long as you are careful not to damage the seeds. It is best to start with a small batch and check to be sure they are not being damaged. Blend a small amount of the seeds in a two to one ratio with water. Use brief, intermittent agitation at low speed and then strain the mixture to separate the seeds from the pulp.

Thrashing seeds (separating seeds from the rest of the collected plant material) is optional, but it does have at least two advantages: it reduces the volume of seeds to be stored, which saves on storage space; and more seed-predators such as insect eggs, mold spores, and other seed disease vectors may be removed with the discarded chaff. The easiest way to thrash seeds is to rub the collected material against a coarse screen with a gloved hand.

Seed Storage

The two most critical necessities for storing seeds are constant temperature and low humidity. A temperature of 50 degrees Fahrenheit or less and 50 percent humidity or lower is ideal. In general, fluctuating temperature and humidity harms seeds more than slightly higher constant values of each.

Store seeds in the refrigerator, not the freezer, until you are ready to plant. Low temperatures, humidity, and darkness protect seed longevity. If it is not practical to store seeds in your refrigerator, store them in any place that is cool, dark, and dry, protecting them from insects as much as possible. Store the seeds in paper sacks to allow good air circulation and prevent molding.

Do not store seeds in plastic bags or other non-breathable containers unless they are air-dried thoroughly first. It is important to include basic information on labels, including date of collection, species name, location of collection, and name of collector.

Dusting the seeds with a mild insecticide will help prevent insect infestation and kill any pests collected with the seeds. Or, you can insert a pest strip for several days while leaving the paper bag open to allow insects to escape. Freezing the seed for a brief period may also be a viable alternative. Seeds of fleshy fruits should be kept moist to maintain viability.

If allowed to dry out, they will either germinate prematurely or not at all. This type of seed should be planted immediately or mixed in a one-to-one ratio of moist sand, sphagnum moss, or a peat and perlite mixture, and stored in a cool place. If the root emerges from the seeds during storage, the seedling should be removed and planted immediately.

Seed storage longevity varies from species to species. Some seeds may be viable after ten years of storage, while others may not germinate after two years in storage. Ideally, seeds should be planted within one year of collection.

SEED CONSERVATION

The implementation of efficient conservation strategies for native plant species is essential to stem the rapid loss of biodiversity throughout Country. 3,500 native vascular plants, over 800 are considered to be threatened in their natural environments, if no actions are taken to secure their future.

The Botanic Gardens of Seed Conservation Centre was established to contribute to the conservation of threatened flora, using ex situ (outside natural environment) methodologies and integrating these collections with the State's in situ (within natural environment) conservation programmes.

The primary objective of the Centre is to undertake seed collections of priority plant species throughout country for the establishment of long-term seed conservation collections and development of germination and storage protocols for each species collected. The Seed Conservation Centre is currently the only unit within country undertaking this type of research with native plant species.

Once a target species is identified in the field, seed is harvested in a manner that ensures a genetic representation of the population is collected. No more than 20% of the available seed is collected; ensuring the long-term impact of seed collection on wild plant populations is minimal. It is however important that sufficient seed be collected, as routine viability and germination trials during the life of the collection use up a proportion of the stock.

Once collected, seed is kept dry in breathable cotton or paper bags until it reaches the Seed Conservation Centre.

Incoming seeds are placed in a drying room, under ambient conditions of 15°C and 15% relative humidity. These conditions ensure slow, steady drying and minimise reductions in seed lot viability. Seeds are then cleaned to remove debris and foreign materials.

For some species it is necessary to extract seed from pods or cones before cleaning can occur. Once cleaned and counted, seeds are returned to the drying room, and left to equilibrate to a water moisture content of 3.5-6.5% depending on oil content within the seed.

Seed Testing

Testing the viability of Whibley's Wattle seed Acacia whibleyana on water agar

A sub-sample of each seed lot is removed for germination trials. Seeds are tested for germination under a variety of conditions on water-agar plates. Species exhibiting dormancy mechanisms are subjected to a range of chemical and physical treatments to promote germination. Optimal conditions are recorded and applied during subsequent viability testing.

SEED STORAGE

A sub-sample of each seed lot is removed for germination trials. Once dried, seeds are transferred to foil pouches, hermetically sealed and stored under low temperature conditions. Although the storage life of seeds is variable, data suggests species producing orthodox (tolerant of desiccation and low temperature storage) seeds can be stored for over 100 years. To test viability and ensure storage conditions are not detrimental to collections, a sub-sample of seed is tested twelve months after first being stored, with subsequent tests every five years.

To determine the suitability of these environmental conditions for conserving seed from native species, the Seed Conservation Centre aims to investigate the effect of storing seeds under low temperature conditions with reduced moisture contents.

Alternative Conservation Techniques

Tissue culture of the endangered Spiny Everlasting Daisy Acanthocladium dockerii.

In some instances, seed from particular plant species may be recalcitrant to long-term storage. Alternatively limitations in population size may prevent sufficient seed being collected from wild populations. Under these circumstances, alternative ex situ conservation techniques including the tissue culture based methods of micro propagation and embryo culture are adopted.

SEED QUALITY

Good seed quality is the result of successful interaction between seed breeding, seed multiplication and seed processing. Breeding selects the required genes. Multiplication ensures that the raw materials are produced under optimum, controlled conditions, whereas processing includes subsequent cleaning, sorting, pelleting and coating of commercial seeds.

Breeding

The sugar content is and will remain generally the most important property in sugar beet breeding, but breeders have

really shifted their focus onto seed quality following the introduction of genetically monogerm (single-germ) seed in the early 1970s. Today, sugar beet seed is a highly specialised commodity, produced by interbreeding genetically monogerm mother plants and pollen-producing father plants.

Genetically, seed quality covers a number of properties. However, the mother plant properties are crucial to seed quality. The mother plant will ensure that seed is monogerm, that seed is properly nourished during the growing process, and that seed possesses the relevant external properties such as size, ability to absorb water, etc. Moreover, the mother plant genes control some of the internal seed properties. These include endosperm growth and hormonal seed balances, to name a few.

When growing new sugar beet varieties that meet the customers' demands for sugar content, juice quality, root shape and resistance properties, Danisco Seed breeders constantly focus on the mother plants. They measure seed yields, seed dormancy, germination capacity and vigour. All parameters of importance to seed quality are recorded, and the overall result determines whether a given mother plant will produce Danisco Seed's commercial seed.

SEED MULTIPLICATION

Monogerm sugar beet seeds are grown in the Po valley of northern Italy and in south-western France. Many years of experience have shown that the best seed quality is obtained in these regions, and both growers and our own technicians have, over the years, built up expertise in growing.

Multiplication of beet seeds is a unique process, and many factors are capable of affecting seed quality. The process spans more than 12 months from August to the following July/ August and consists of two stages. The first stage is the production of cuttings, beginning with the sowing of basic seeds in August and ending in February with vernalised (affected by cold) cutting of beets followed by planting in the propagation field. Germination, establishment, irrigation, fertilisation and not least winter temperatures matter to the number and quality of cuttings.

The second stage is the propagation field. The beet cuttings are harvested in early February. They are delivered directly to the growers, where multiplication takes place. Here, they are sorted according to size and planted in the ratio of six rows of MS (mother plants) to two rows of pollinators (father plants). In March, April and May, they grow, bolter and flower from late May. The flowering process continues for nearly a month until mid-June, and the seeds develop and grow the following months. During the entire growth season, our own technicians monitor the fields closely. Once a

week, the fields are inspected, and growers receive instructions on irrigation, fertilisation, disease and pest control, and - before seed flowering - topping and synchronisation. The purpose of topping is to make MS plants uniform with the most developed plants being pruned. The objective of synchronisation is to ensure that MS and pollinator plants flower at the same time. If one component has grown more than the other, efforts must be made to 'delay' it through pruning. Punctuality in all processes is vital to seed growth and quality.

From mid-July, the seed plants will pass the peak of ripening and become ready for swathing. Finding the right time for swathing, i.e. the time when the highest number of seeds is ripe, is very critical. Consequently, our workers inspect the fields frequently during this period and agree on the exact swathing time with the growers. Early plant swathing will slow down the germination speed of the finished product. Late swathing will affect yields!

The seeds are threshed after 3-5 days of swathing. The seeds are subsequently pre-cleaned and sealed before being shipped off to Danisco Seed's seed factory at Holeby in Denmark, where further processing takes place.

SEED PROCESSING

During the first stage of seed processing, seeds are cleaned, polished and calibrated. Each individual batch of seeds received from the growers is processed separately, and we analyse samples in respect of size and quality before the cleaning process starts so that we know what to expect and thus can set our machinery correctly. During the process, quality is controlled regularly, and machinery is adjusted accordingly.

Genetics (mother plants) affects the amount of polishing to be carried out in respect of seeds. Polishing is required to remove excess cork and agents inhibiting germination. Some mother plants produce seeds with a thick cork layer, whereas other seeds have a thinner

cork layer. Seeds with a thin layer of cork are easily bruised during the process if care is not taken. Consequently, seed processing is really manual labour, where the skills, material knowledge and care of our technicians are critical to achieving a good result.

At the next stage, seeds must be pelleted. A mixture of water, glue and wood dust makes the seeds round so that they fit into the growers' seed drills. The wet pellets are calibrated into the right size and are subsequently allowed to dry.

Different fungicides will be added to the pellet mass, depending on the market for which seeds are intended.

Finally, pesticides are added to the pellet surface and the pellets are then coated with a coloured film preventing users from getting into physical contact with the chemicals. Danisco's characteristic green colour makes it easy to spot the seeds in the soil when sowing depth and sowing distance are checked.

Because beet seeds are sown for permanent plantation, it is important that fungicides and insecticides, are distributed evenly to the individual pellets since too low doses will provide inadequate protection against pests, whereas too high doses may inhibit germination.

To ensure a high, uniform quality, only 25-30% of the original raw materials will end up as commercial seeds. The remaining quantities are sorted out during the process and burnt.

Chapter 8

Hybrid Seeds

Hybrid varieties have been evolved in those high valued vegetable crops which exhibit marked heterosis such as solanaceous vegetables (tomato, eggplant, sweet pepper), cucurbits (melons, watermelon, cucumber, squash, pumpkin and gourds), cole crops (cabbage and cauliflower), root and bulb crops (onion, radish, carrot) and fruit vegetable like okra.

SOLANACEOUS CROPS

Tomato

Manifestation of Heterosis

Hybrid vigour in tomato is manifested in the from of earliness, total yield, uniformity of produce, greater plant vigour and better adaptability to unfavourable environment. In general heterosis for yield in hybrids ranged from 30-50 percent.

Hybridization Techniques

For commercial hybrid seed production the emasculation of female flowers and pollination by hand is still considered economical and efficient. Most of the seed companies do hybridization this way.

Indeterminate tomato varieties are staked and trained with either single stem or double stem, whereas, the determinate tomato varieties are trained with 3 stems. Usually 1st to 4th cluster on each branch are selected for emasculation.

Buds where the corolla leaves have just opened and form an angle up to 45" in respect to the flower axis (a day prior to anthesis) are selected for emasculation.

The emasculation operation involves in holding the corolla at the base and with a single upward pull pick off the corolla along with all the stamens. Usually the anthers are picked off a day before anthesis with the help of forceps leaving the petals intact.

Such petals turn yellow on the day of anthesis. Fresh pollen collection on the day of anthesis by a vibrator has the highest viability, because only ripe pollen are shed by vibrating the flowers. Pollens are collected in a glass tube or on a glass plate from the male line and are transferred to the stigma by finger or by inserting the stigma into a glass tube containing the pollen grains.

Left over pollen grains in the glass tube are not used on the next day because its viability is reduced considerably. Half of the calyx of pollinated flowers is removed to distinguish it from unpollinated flowers. Pollen grains can be stored for a longer period (2 months) of time when its moisture content is reduced using a desiccator and the temperature is kept around 0°C.

Physiological Parameters Affecting Tomato Seed Yield and Vigour

Growth of tomato plant is satisfactory upto 25°C. In tropical regions, prolonged high temperatures (35-42°C) adversely affect pollen fertility and physiology of fertilization, of inflorescence, different altitude and latitude and fertilizer application exert an influence leading to poor seed set. Age of pollen and stigma, abundant or scarce pollination, height on the vigour of the plant.

Improvement in Methods of Hybrid seed Production

Nearly 40 percent of the total labour expenditure is on flower emasculation during the course of hybrid seed production, which can be reduced by using male sterile lines.

Male sterility controlled by recessive genes (totaling 42 in number now) and stamenless mutant controlled by recessive gene (s1) having normal corolla and free stigma, are quite accessible for pollen application.

Limited success has also been obtained in phenotypical restoration of sterility in 'ms' forms with silver nitrate treatment. Partial success has been achieved in restoration of (ms) sterility using GA 3 or GA 4/7. Such restoration will result in the production of 100% fertile plants from male sterile forms.

Seed Extraction and Drying

Tomato seeds are extracted mainly by fermentation method. Under warm conditions, the fermentation process is complete in 24 hours. At 25ºC, it requires 2 days for the completion of fermentation process. The pulp is stirred several times in a day to maintain a uniform rate of fermentation and to avoid discolouration of the seed. Seeds are than washed with clean water.

Fermentation method of seed extraction also controls bacterial canker disease which is seed born. Tomato seed is also extracted using acid (HCl) or alkali (NaOH), 10 cc or 36 percent HCI or 30 percent NaOH is added in 4 kg of tomato pulp. The treatment is given for a period of 15 minutes, which separates the jelly from tomato seeds. The seeds are than washed thoroughly and then dried.

Seed Yield

One kg of tomato fruit will produce 3-4 g of seed yield (1000-1200 seed). Av. seed yield : 60-70 kg/ha depending upon the performance of parental lines.

EGGPLANT AND SWEET PEPPER

Manifestation of Heterosis

In general heterosis in sweet pepper ranged from 35-40 percent whereas, in eggplant it ranges from 50-150 percent. Rashid et al. reported 50% heterosis in eggplant in Bangladesh. Heterosis in eggplant is manifested in earliness, fruit number per plant and fruit weight. In case of sweet pepper, heterosis is manifested for plant height, days to flower, fruit weight, number of fruits per plant, early and total yield.

Hybridization Techniques

Emasculation and hand pollination is the useful production techniques. Stigma is receptive a day prior to anthesis in eggplant. Hence bud pollination is possible giving good fruit set and seed yield. In sweet pepper, emasculation is done a day prior to anthesis, whereas, pollination is done in the morning on the day of anthesis. Natural cross pollination ranged from 0.2-46.8% in eggplant flowers.

Emasculated flowers are never visited by pollinators. Pepper flowers are visited by honey bees occasionally. Fresh pollen grains are collected on the day of anthesis by a vibrator and can be stored for a period of 1 to 2 months at *0ºC*, using silica gel for proper drying of the pollen grains.

In order to obtain optimum yield and good quality seed, it is essential to train the eggplant as well as sweet pepper plants. The first and 2nd flowers are harvested at the initial stages. This will boost plant growth as well as the number of seeds formed in subsequent fruits. The training of eggplant involves to allow two lateral branches below the first flower and the remaining lateral branches are removed. This technique is aimed to attain sound growth of plant.

Physiological Parameters Affecting Seed Yield and Vigour in Eggplant and Sweet Pepper

The eggplant is photoinsensitive but requires an optimum day temperature of 25-30°C and 20-27°C night temperature for its growth. In case of pepper effect of low temperature (8-10°C) on seed set is greater before anthesis than afterwards.

Plant growth is satisfactory at day temperature ranging from 20-25°C. Pepper plant prefers high humidity for its proper growth. Providing partial shade prior to flowering boost its vegetative growth. Capsicum flowering and fruit set require bright sunlight.

Use of Male Sterility in Hybrid Seed Production in Eggplant and Sweet Pepper

Male sterile lines are available in eggplant (genic male sterility) and sweet pepper (genic and cms) but are not successful at commercial seed production. Seed set on male sterile lines ranged from 46-67% in capsicum. Use of functional male sterility in hybrid seed production, sweet pepper has a bright future. Cytoplasmic male sterility is being used now in case of chilli pepper to produce F1 hybrid commercially by several seed companies.

Seed Extraction

Eggplant fruits are harvested 50-55 days after anthesis and are stored for a period of 10 days for postharvest repening. Sweet pepper fruits are harvested 60-65 days after anthesis. The ripe fruits are crushed and seeds are separated by washing with excess of water without fermentation. The seeds are dried using dry air at 28-30°C.

Seed Yield

A satisfactory seed yield in eggplant is 150-200 kg/ha with a thousand seed weight ranging from 4-5 g. In peppers the seed yield varies from 100 to 200 kg/ha. One kg of sweet pepper will yield 5-7 g of seed with the thousand seed weight equals to 5 g.

CUCURBITACEOUS CROPS

Sex Expression and Sex Forms in Cucurbits: Cucurbitaceous plants produce three types flowers depending on species and variety. One plant my contain more than one type of flowers, and more than one sex forms may be available in a species. According to the sex forms, cucurbitaceous plants can be grouped as follows:

- *Hermaphrodite*: All flowers of a plant are bisexual (available in some varieties of Luffa, Cucumis and Benincasa spp.)

- *Monoecious:* Male and female sexes are in different flowers of the same plant (the most common sex form).
- *Andromonoecious:* Male and hermaphrodite flowers are in the same plant.
- *Gynomonoecious:* Female and hermaphrodite flowers are in the same plant.
- *Trimonoecious or Androgynomonoecious:* Male, female and hermaphrodite flowers are in the same plant.
- *Dioecious:* Male and female flowers are in different plants. Plants bearing male flowers called androecious and those bearing female flowers called gynoecious plants.
- *Gynodioecious:* In dioecious species some plants of a variety bear only female flowers and others bear hermaphrodite flowers.
- *Sub-gynoecious:* When gynoecious plants of dioecious species produce some male or hermaphrodite flowers.

Monoecious sex form is the most common one in this family. Hermaphroditism is the original sex form from which monoecious and dioecious sex forms have been evolved.

Pollination and Fruit Set

Dioecious and monoecious plants are cross pollinated. Hermaphrodite flowers also fail to set seed by self-pollination some times as the pollen grains are sticky. However, self-pollination between two sexes of the same plant is also common in cucurbits. Poor fruit setting in cucurbits is primarily assumed as the failure of pollination.

However, this situation may be improved by hand pollination. Fruit setting is largely dependent on varietal characteristics, nutrition, disease status and environment. It also depends on the number of fruits already present in a plant. Therefore, harvesting of edible-mature fruits encourages new fruit setting.

Methods of Hybrid Seed Production in Cucurbitaceous Vegetables

Most of the cucurbits are cross-pollinated because of their monoecious flowering behaviour. However, inbreeding may take place without any barrier as both the sexes are present in the monoecious sex form. Experimental results have demonstrated that cucurbits as a whole exhibit no inbreeding depression. It should be pointed out that there are twopopular cucurbitaceous vegetables-Kakrol and pointed gourd (potal), which are of dioecious flowering habit leading to 100% cross-pollination.

These crops may have tremendous inbreeding depression, especially in Kokrol even after one generation of selfing. Inbreeding in dioecious species is not possible in conventional methods as the male and female sexes are in different plants. Moreover, time of anthesis, time and duration of stigma receptivity and pollen availability are different in different cucurbits. Therefore, hybrid seed production technique for each of the cucurbits has been discussed separately.

Different steps of Hybrid Seed Production are:

- Production of inbred lines by inbreeding for 3 to 5 generations.
- Selection of inbred parents through combining ability tests and potential hybrid production ability.
- Production of hybrid seeds (preferably single cross hybrids and pistilate parent preparation is relatively easy, and single fruit produces quite a large number of seeds),
- Maintenance of inbred parents.

WATERMELON

Inbreeding and Selection

Inbreeding may be defined as any system of mating that will lead to an increase in homozygosity. In watermelon, inbreeding is usually practiced by controlled self-pollination.

This provides the most effective approach to obtain homozygosity. Inbreeding in watermelon does not cause loss of vigour, self-sterility, decrease in the number of fruits or total yield per plant.

Porter stated that inbreeding tends to isolate strains which produce either larger or smaller fruits than the commercial variety. It was also reported by Porter that in Klondike variety, inbred for four successive generations have been isolated which produce fruits equal in weight to those of the parent variety but excel the latter in fruit uniformity, flesh colour, texture and quality. Single plant selection and compositing the seeds of selected plant after five successive generation will lead the inbred development in watermelon.

Technique of Hybrid Seed Production

Hybrid seed production in watermelon entails little difficulty because of the monoecious nature of the flowers and the large numbers of the seeds contained in a fruit.

Selection of Parental Line

For hybrid seed production, it is very important to choose the parental line for cross combination. The selection of parent should be based on combining ability, cross compatibilitys tudy and hybrid vigour manifested in F1 generation.

At BARI two parental combination viz. WM0024 x WM0045 and WM0053 x WM0045 were evaluated based on heterotic performance and cross compatibility and one of the combination was released as F1 PADMA for its light green colour, oblong shape, high sugar percentage and better overall performance.

Maintenance of Parental Line

When a new superior combination of parent lines for an hybrid variety is identified, the first step should be taken on the process of seed production of parent lines by selfing. For the production of seeds of inbreds the following points should be avoided - i) to produce seeds of parent lines every year and

ii) to improve the genetical characters of the parent lines by means of producing their seeds early.

Hybrid seeds of watermelon can be produced by two ways.

1. *Through artificial pollination.*
2. Removal of maleflower and use of insect pollination.

Artificial Pollination

Field Lay-out:

Seedlings at 3-4 true leaf stage should be transplanted at 1.5 m apart in 2.5 m wide beds. At transplanting time, male parent line should be transplanted in the separate rows, so that they never get mixed with female parent lines.

Selection of Female Flowers

Since watermelon plant bears a number of ineffective female flowers which may fail in fruit set, therefore, effective female flowers have to be selected in order to make successful pollination. Strong and stout female flowers having large ovary with long peduncle are best for effective pollination. The female flower bud which will open next day should be selected for bagging.

Bagging of Female and Male Flowers

The process of bagging are as follows.

- Find out adequate female flowers
- Put the bag covering on the bud
- Insert the bud with vine up to the end of bag and
- Fix the bag on the vine by means of twisting 2-3 times the cut ends of the bag at opposite of the vine

Collection of Male Flowers

There are two methods of collection of male flowers. One is that considering convenience of pollination, male flower buds can be collected in the late afternoon of previous day, keeping their peduncles as long as possible and store them up to the next morning. Another method is to collect the male flowers in early morning before dehiscence of anthers.

Period *of* Pollination

Main pollination period is usually 7 days, 2 days for beginning, 2 days for the peak and another 2-3 days for the last. The peak period of flowering occurs almost in the same dates, therefore, if needed to widen the pollination period, the design of planting date must be adjusted at 15 days interval.

Correlation Between Pollination Hours and Fruit Set

From the mid February to the end of March, flower buds of watermelon starts to open from *6:00* A. M, fully blooms by 700 A. M and completely closes by 11:30 A. M. *On* the other hand during the end of June to the beginning of July, flower bud open from 4:30 A. M, fully blooms by 5:30 A. M and completely closes by 10 A. M.

Pollination

- Collect the male flowers in the box or petri dish.
- Pick the male flower from the box and break the petal, then hold it by the mouth at its peduncle.
- Remove the bag carefully from the opened flower.
- Pick the male flower, hold it at its peduncle and put its anther mass on the pistil tapping all the divided 3-4 stigmas evenly and wholly to pollinate completely.
- Hold the male flower again by the mouth, then put bag on the pollinated flower.
- Put the mark-tag on the vine at the female flowers internode.

In the case of rainy days, the pollinated flowers should have to be protected from rain. Watermelon pollens are very susceptible to water and if touched with water drop the pollens are destroyed immediately.

SEED FRUIT MANAGEMENT DURING GROWTH PERIOD

When the fruits become the size of baby's head, spread the straws under the fruits in order to avoid phytophthora rots.

Harvesting

Fruits of 30-35 days old from the date of pollination, become ready for harvesting. After harvesting, keep the fruits in well ventilated room for the post-harvest ripening in order to improve the seed colour and for the convenience of seed extraction. When the fruits become over ripe, the seed extraction becomes very easy due to softening of flesh.

Extraction, Washing and Drying of Seeds

It is wise to complete the seed extraction and drying on the same day, therefore, referring the weather forecast and choosing fine day the work is to be started from early morning. The process is as follows. Seeds are mixed with pulp or placenta in cucumber, watermelon etc. Fruits are cut longitudinally into half and scraped out the seed and collected in a barrel. Some placenta remains with the seeds which is to be removed by rubbing with sand or ash followed by washing in water.

Otherwise, the pulp surrounding the seeds can be allowed to ferment for 48 hours, when the pulp can be easily separated and then seeds are washed in fresh water. A rapid method of separating the seed from pulp is by acid treatment. Twenty five to thirty ml of hydrochloric acid or about 8 to 10 ml of commercial sulfuric acid can be used for 5 kg of pulp containing seeds. The seeds can be washed free from the pulp in about 20 to 30 minutes. Then the seeds have to be washed thoroughly to remove excess acid and undeveloped floating seeds are discarded.

Seed yield per hectare varies from 150 kg to 300 kg which may vary based on varieties, extent of pollination and field condition.

Removal of male flowers and Use of Insect Pollination

The technique is applicable to monoecious species of the cucurbits, in which male and female parental line should be planted in alternate rows. The male flowers of female plant should be completely removed before their opening. The fruits from the female parent are harvested as crossed fruits and those of the other variety as self fruit of the male variety.

This technique has been used in commercial hybrid seed production.In Canada hybrid seeds of muskmelon, cucumber, squash etc. are produced by planting one row of male parent plant and two rows of female parent plants alternately. This methods is referred to as crossing block method. In Japan three workers are required for 0.245 acres to carry out their operation.

PUMPKIN

Production of Inbred Lines

- Seed sowing of diversified genotypes in November to January for winter ecotypes giving 10.6 m distance within and between lines. For summer ecotypes, seed sowing preferably after the 1st monsoon. Allow the plants to climb on narrow bamboo trellis or vertical net-trellis giving 1.0 m interval between plants.
- Bagging of male and female flowers one day before anthesis, and also after pollination of the pistilate flowers for another two days.
- Select the vigorous inbreds after 4-5 generations of inbreeding.

Production of Hybrid Seeds

- Planting of female and male inbred parents in 4 : 1 ratio,
- Spray 50 to 100 ml of ethephon per litre of water at 2-3 leaf stage on pistilate parent to increase female flowers at lower nodes.

- Allow the plants to climb on narrow or vertical net-trellis.
- Bagging of male and female flowers every afternoon before anthesis. Rebag the female flowers after pollination for another two days.
- Harvest fruits after about 60 days of pollination.
- Remove seeds from the fruits, wash and dry at low temperature (<30ºC) for 3-4 days and then sun dry for another 3-4 days.
- Preserve the seeds in sealed polyethylene bags at low temperature (4-5°C).

Maintenance of the Inbred Parents

Inbreeding of parents to produce seeds provided the parental stock is depleted. Produce large quantity of seeds to maintain genotypic and phenotypic integrity of the hybrid. Seed Extraction : As in watermelon.

Bottle gourd

In cucurbits, bottle gourd (2n = 22) is second to the pumpkin by area (7300 ha) of cultivation and total production (60,000 t/ year) in Bangladesh. Probable centre of origin is Africa.

Production of Inbred Lines of Diverse Genetic Background

Bagging of male and female flowers before anthesis

Inbreeding at anthesis through following morning and rebagging of the female flowers for another 2 to 3 days.

Inbreeding should be done for 4-5 generations and select the better types by discarding the poor performers.

Production of Hybrid Seeds

Planting of female and pollinator inbreds in 4:1 ratio.

Bagging of female and male flowers before anthesis, and only female flowers after pollination for another 2 days.

Harvesting of mature fruits after senescence of the plant, remove seeds, wash and dry them. Store in sealed polyethylene bags at low temperature.

Maintenance of The Inbred Parents

Inbreeding of parents to produce seeds should be done when the parental stock is depleted. Produce large quantity of seeds to maintain genotypic and phenotypic integrity of the hybrid.

WHITE GOURD

Area of white gourd ($2n = 22$) cultivation in Bangladesh is 3,600 ha and total production is 22,000 t/year. Probable centre of origin is Java, Indonesia. Av. yield = 6.11 t/ha.

Production of Hybrid Seeds

Planting of female and pollinator inbreds in 4: 1 ratio.

Vertical net-trellis should be allowed for climbing.

Bag the female and male flower buds before anthesis and rebag the female flowers after pollination for another 2 days.

Harvest mature fruits after 70 to 90 days depending on inbreds. Remove seeds, wash, clean and dry them. Store in sealed polyethylene bag at low temperature.

RIBBED GOURD

Total area under ribbed gourd (2n = 26) cultivation is 4000 ha and total production is 17,000 t/year in Bangladesh. Indian subcontinent is the centre of origin. Av. yield is 4.25 t/ha.

Methods of Hybrid Seed Production

Production of Gynoecious and Monoecious and Moecious/ Harmaphrodite Inbred Lines Having Distant Genetic Background. Gynoecious plants should be treated with 300-400 ppm GA3/AgNo3 to induce male sex for selfing to produce gynoecious inbred.

Bagging of male and female flowers in monoecious plants before anthesis. Pollination in the afternoon and rebagging of the female flowers for another 2 days. Inbreeding should be done for 4-5 generations and select the better types by discarding the poor performers. Selection of male and female inbred parents on the basis of their hybrid performances and considering their SCA and GCA value.

Production of Hybrid Seeds

Planting of gynoecious/gynoedioecious/monoecious inbred as female, and monoecious/hermaphrodite/ andromonoecious as male in 4:1 ratio.

Remove hermaphrodite plants from gynodioecious female inbred parent retaining the gynoecious plants only, OR in case of monoecious parents, bag the male and female flowers before pollination and rebag female flowers for another 2 days after pollination.

Harvest mature fruits and dry them before seed removal. Seeds should not be stored at low temperature.

Maintenance of The Inbreds

Gynoecious line should be maintained by inducing male sex using 300 - 400 ppm GA3 or AgNo3.

Hermaphrodite inbreds can be maintained by simple selfing, and for other sex forms inbred can be maintained in the similar way mentioned for other monoecious species.

SPONGE GOURD

Area of cultivation and total production of sponge gourd (2n=26) in Bangladesh are not known. However, it becomes available in lean period of vegetable supply. Centre of origin is India.

Objectives

High fruit yield having non-bitter taste

Early fruiting habit having late fibre formation in fruit

Resistance against CMV, mites, powdery mildew and downy mildew.

Floral Biology

Anthesis at late night

Another dehiscence in the morning with the rise of temperature.

Stigma becomes receptive at anthesis and remains so until noon.

Production of Hybrid Seeds

Planting of female and pollinator inbreds in 4: 1 ratio.

Vertical net-trellis should be allowed for climbing.

Bag the female and male flower buds before anthesis and rebag the female flowers after pollination for another 2 days.

Harvest mature fruits after 70 to 90 days depending on inbreds. Remove seeds, wash, clean and dry them. Store in sealed polyethylene bag at low temperature.

SNAKE GOURD

Area of snake gourd ($2n = 22$) cultivation is 2000 ha and total production is 9000 t/year in Bangladesh. Centre of origin is India.

Objectives

High yield with thick mesocarp of the fruits.

Free from bitterness.

Variety with early and late seeding potential.

Floral Biology

Anthesis at or after sunset.

Stigma becomes receptive at anthesis and remains so far at least another 12 hours.

Pollination can be done in the following morning of anthesis.

Production of Hybrid Seeds

Planting of female and pollinator inbreds in 4: 1 ratio.

Vertical net-trellis should be allowed for climbing.

Bag the female and male flower buds before anthesis and rebag the female flowers after pollination for another 2 days.

Harvest mature fruits after 70 to 90 days depending on inbreds. Remove seeds, wash, clean and dry them. Store in sealed polyethylene bag at low temperature.

BITTER GOURD

Area of bitter gourd (2n = 22) production is 4000 ha and total yield is 15,000 t/year in Bangladesh. Centre of origin may be Asia, Africa or tropical America. Av. yield = 3.75 t/ha.

Objectives

High yield with at least moderate size of fruit in winter months and long fruits in summer season.

Less bitter taste with thick mesocarp having poor number of small seeds.

Floral Biology

Anthesis at early morning.

Stigma becomes receptive at anthesis and it remains so for another 5-6 hr.

Pollen dehiscence starts at anthesis and it increases with the rise of temperature.

Production of Hybrid Seeds

Planting of female and pollinator inbreds in 4: 1 ratio.

Vertical net-trellis should be allowed for climbing.

Bag the female and male flower buds before anthesis and rebag the female flowers after pollination for another 2 days.

Harvest mature fruits after 70 to 90 days depending on inbreds. Remove seeds, wash, clean and dry them. Store in sealed polyethylene bag at low temperature.

KAKROL OR TEASLE GOURD

Exact area of cultivation and total yield is not known. However, large area of cultivation is seen in Brahmanbaria, Akhaura, Rangpur and some other areas. Kakrol (2n = 56) has export potential to Middle East and U. K. Centre of origin probably is Indo-Burma region. It is generally a vegetatively propagated dioecious crop.

Objective

High yield with big size of fruits.

Reduced seed number in fruits, and poor root tuberization.

Elimination of cumbersome hand pollination practice.

Floral Biology

Anthesis at early morning.

Stigma becomes receptive at anthesis and receptivity remains until mid day.

Pollen becomes viable with the rise of temperature, at about 6:00 to 700 A. M.

Method of Hybrid Seed Production

Production of Inbreds of Different Clones on The Basis of Fruit Characteristics

Grow different fruit morphotypes from tubers on vertical trellis of net giving 0.8 m distance between plants in a row.

Treat some of the twigs of each clone with 400 ppm AgNo3 before flower bud organogenesis to induce male sex in female plants. Treatment should be given by spraying the solution on top 7-8 leaves of the twigs.

Cover the flowers of treated and untreated vines one day before anthesis.

Pollinate the stigma of untreated vines with induced pollen of the same plant or different plants of the same clone. Rebag the pollinated flowers for 3 days to avoid genetic contamination.

Collect ripe fruits and dry the seeds for storing at low temperature.

Deshelled seeds, without any injury in the embryo, should be sown in moist sterile medium (sand, soil, vermiculite, etc.) maintaining at least 30°C for about 10 days. Seedlings should be established in February every year.

Selection of Inbred Parents

Select two genetically female inbreds, as male and female parents of the hybrid, on the basis of their hybrid performances, SCA and GCA value. Male sex must be induced while producing their hybrids.

Production of Hybrid Seeds

Transplant tuber-originated inbred parents side by side, and allow them to grow on vertical trellis so that pollination can be done conveniently.

Induce male sex in at least one of the parents (pollen parent) by spraying AgNo3 (400 ppm) in twigs before flower bud organogenesis.

Bag the flowers of seed and pollen parents 1 day before anthesis.

Pollinate the flowers by induced pollen and rebag the pistilate flowers for 3 days.

Harvest the hybrid seeds at full maturity and dry them for storing at low temperature.

Distribute the hybrid seeds directly or establish the hybrid tubers from F1 seeds and distribute them for commercial production with natural male in 10:1 ratio (female : male) provided there is no parthenocarpy.

Maintenance of The Inbred Parents

Just maintain the seed-originated tubers of the inbred lines and multiply according to necessity.

Inbreeding depression may be very high in kakrol.

Chance hybrids, though relatively low, from ordinary crosses of male and female genotypes may also be selected and multiplied as variety.

Similar methods of hybrid seed production may be applicable in pointed gourd. However, seed-originated plants take several years for fruit production. Therefore, generation advancement may take time during inbred production.

ONION

Heterosis in Onion

Heterosis in yield of onion hybrids ranged from 14 to 67 percent when compared with commercially grown onion varieties. Heterosis is manifested in uniform bulb size, bulb weight and an efficient source sink ratio.

Male Sterility in Onion

Male sterility in onion is controlled by the combination of a cytoplasmic factor 'S' together with a recessive nucleus gene in its homozygous form (ms). These male sterile lines are maintained by pollination with a maintainer line of construction 'N ms ms'.

Hybrid Seed Production Techniques

For the production of hybrid seed in onion, the male and cytoplasmic female lines are planted in the ratio of 2% The success of hybrid seed production depends upon the pollen distribution pattern from fertile to sterile plants in the crossing block.

Factors Affecting Hybrid Seeds in Onion

The following factors are responsible for reduced hybrid seed yield in onion.

Weak Inbred lines

Aborteds ovule

Abnormal florets where ovary started to develop but failed to produce seed.

Asynchrony of flowering of parental lines

Excessive heat damaging the flowers.

Seed Yield in Onion can be Improved as Follows:

- Better synchrony of flowering of parental inbreds is achieved by adjusting either storage temperature of the mother bulb (9-14°C) or by planting dates.
- Keeping bee colonies in the hybrid seed production plots @ 3-5 dully developed bee colonies /ha.
- A single application of GA at 50 ppm at the time of first seed stalk emergence reduces the times of 80% of floral stem emergence by half and improves the uniformity of seed stalk height.
- Harvesting seed having 60-70% dry matter content while still in capsule attached to the stalk.
- Avoid shattering of seeds by spraying antishattering materials such as polyvinyl acetate.
- Spray desiccants (diquats) to facilitate uniform drying and mechanical harvesting.

Seed Yields

Hybrid seed yield in onion ranges from 300-350 kg per hectare.

COLE CROPS

Manifestation of Heterosis

Heterosis in cabbage (25-61%), cauliflower (20-60%) and Broccoli (26-58%) is manifested in head/curd size, early maturity, head/curd weight, and plant weight.

Incompatibility and Hybrid Seed Production in Cole Crops

Mild winters (0-5°C) with spring temperature at bloom (15-20°C) are very suitable for cole crop seed production. In cole crops it is the sporophytic incompatibility system which is most prevalent.

The self- and sib-incompatible but cross-compatible lines are set in the field with a planting ratio of one row of pollinator to four rows of seed parent. The lines A (S11) and B (S22) are each propagated by bud pollination or through tissue culture. The hybrid seeds (S1 S2) are harvested from both parents. Such hybrids are the most uniform ones but usually the production cost of the parental lines is prohibitive.

Main problems in hybrid seed production using incompatibility system are:

1. Depression by continuous inbreeding of parental lines
2. Pseudo compatibility
3. Reduction of incompatibility by environmental conditions
4. Restriction of pollination within parental lines and
5. Matroliny.

Male Sterility and Hybrid Seed Production

Cytoplasmic male sterile lines and their maintainers have been developed in cauliflower, cabbage and broccoli using radish cytoplasm for male sterility. Cytosterile plants of broccoli, cabbage and cauliflower are petaloid with large nectaries responsible for bee attraction and good female fertility.

Seed Yield

It varies from 500 - 800 kg/ha3

Chapter 9

Seed Born Diseases

Early blight is a very common disease of both potato and tomato. It causes leaf spots and tuber blight on potato, and leaf spots, fruit rot and stem lesions on tomato. The disease can occur over a wide range of climatic conditions and can be very destructive if left uncontrolled, often resulting in complete defoliation of plants. In contrast to the name, it rarely develops early, but usually appears on mature foliage.

Symptoms

On leaves of both crops, the first symptoms usually appear on older leaves and consist of small, irregular, dark brown to black, dead spots ranging in size from a pinpoint to 1/2 inch in diameter. As the spots enlarge, concentric rings may form as a result of irregular growth patterns by the organism in the leaf tissue. This gives the lesion a characteristic "target-spot" or "bull's eye" appearance. There is often a narrow, yellow halo around each spot and lesions are usually bordered by veins. When spots are numerous, they may grow together, causing infected leaves to turn yellow and die. Usually the oldest leaves become infected first and they dry up and drop from the plant as the disease progresses up the main stem.

On tomato, stem infections can occur at any age resulting in small, dark, slightly sunken areas that enlarge to form circular or elongated spots with lighter-coloured centres. Concentric markings, similar to those on leaves, often develop on stem lesions. If infested seed are used to start tomato transplants, seedlings may damp off soon after emergence. When large lesions develop at the ground line on stems of transplants or seedlings, the plants may become girdled, a condition known as "collar rot." Such plants may die when set in the field or, if stems are weakened, may break over early in the season. Some plants may survive with reduced root systems if portions of stems above the canker develop roots where they contact the soil. Such plants, however, usually produce few or no fruits. Stem lesions are much less common and destructive on potato.

Blossom drop and spotting of fruit stems, along with loss of young fruit, may occur when early blight attacks tomatoes in the flowering stage. On older fruits, early blight causes dark, leathery sunken spots, usually at the point of stem attachment. These spots may enlarge to involve the entire upper portion of the fruit, often showing concentric markings like those on leaves. Affected areas may be covered with velvety black masses of spores. Fruits can also be infected in the green or ripe stage through growth cracks and other wounds. Infected fruits often drop before they reach maturity.

On potato tubers, early blight results in surface lesions that appear a little darker than adjacent healthy skin. Lesions are usually slightly sunken, circular or irregular, and vary in size up to 3/4 inch in diameter. There is usually a well defined and sometimes slightly raised margin between healthy and diseased tissue. Internally, the tissue shows a brown to black corky, dry rot, usually not more than 1/4 to 3/8 inch deep.

Causal Organism

Early blight is caused by the fungus, Alternaria solani, which survives in infected leaf or stem tissues on or in the soil. This fungus is universally present in fields where these crops have been grown. It can also be carried on tomato seed and in potato tubers. Spores form on infested plant debris at the soil surface or on active lesions over a fairly wide temperature range, especially under alternating wet and dry conditions.

They are easily carried by air currents, windblown soil, splashing rain, and irrigation water. Infection of susceptible leaf or stem tissues occurs in warm, humid weather with heavy dews or rain. Early blight can develop quite rapidly in mid to late season and is more severe when plants are stressed by poor nutrition, drought, or other pests. Infection of potato tubers occurs through natural openings on the skin or through injuries. Tubers may come in contact with spores during harvest and lesions may continue to develop in storage.

1. Use a crop rotation that includes potatoes or tomatoes only every third or fourth year to allow infested plant debris to decompose in the soil. Rotations with small grains, corn or legumes are preferable.
2. Use tillage practices such as fall plowing that bury all plant refuse.
3. Select cultivars that have a lower susceptibility to early blight.
4. Use certified disease-free tomato seed and transplants.
5. If producing tomato transplants, disinfest soil in plant beds and control humidity in cold frames or greenhouses. Practice good sanitation throughout the transplant production operation.
6. Use appropriate measures to control weeds and volunteer potatoes and tomatoes in production areas.
7. Maintain fertility at optimal levels-nitrogen and phosphorus deficiency can increase susceptibility to early blight.
8. Time applications of overhead irrigation to allow plants to dry before nightfall.
9. Although the above measures are important to minimize infection, it is usually necessary to apply fungicide sprays to fully protect plants from early blight. Applications to tomato are usually begun 2-3 weeks following emergence or soon after transplanting if a calendar schedule is followed.

Timing of fungicides can also be made using the TOM-CAST system. For potatoes, fungicide applications should be initiated when plants begin to flower. Thorough coverage is important. Adjust equipment for good vine penetration.

Late Blight of Potato and Tomato

Late blight is one of the most devastating diseases of potato and tomato worldwide. It was responsible for the devastating Irish potato famine of the 1840's and has continued to be important to the present. Since 1990, late blight has caused widespread damage across the United States and Canada. If left unmanaged, this disease can result in complete destruction of potato or tomato crops.

Symptoms

Late blight appears on potato or tomato leaves as pale green, water-soaked spots, often beginning at leaf tips or edges. The circular or irregular leaf lesions are often surrounded by a pale yellowish-green border that merges with healthy tissue. Lesions enlarge rapidly and turn dark brown to purplish-black. During periods of high humidity and leaf wetness, a cottony, white mold growth is usually visible on lower leaf surfaces at the edges of lesions. In dry weather, infected leaf tissues quickly dry up and the white mold growth disappears. Infected areas on stems appear brown to black and entire vines may be killed in a short time when moist weather persists.

On potato tubers, late blight appears as a shallow, coppery-brown, dry rot that spreads irregularly from the surface through the outer 1/8-1/2 inch or more of tissue. On tuber surfaces, lesions appear brown, dry, and sunken, while infected tissues immediately beneath the skin appear granular and tan to copper-brown. When tubers are stored under cool, dry conditions, lesion development is retarded and, upon prolonged storage, lesions may become slightly sunken and desiccated. Secondary bacteria and fungi frequently enter late-blight lesions, usually resulting in a slimy breakdown of entire tubers.

Late blight can also develop on green tomato fruit, resulting in large, firm, brown, leathery-appearing lesions, often concentrated on the sides or upper fruit surfaces. If conditions remain moist, abundant white mold growth will develop on the lesions and secondary soft-rot bacteria may follow, resulting in a slimy, wet rot of the entire fruit.

Causal Organism

Late blight is caused by the fungus *Phytophthora infestans*. Unlike most pathogenic fungi, the late blight fungus cannot survive in soil or dead plant debris. For an epidemic to begin in any one area, the fungus must survive the winter in potato tubers (culls, volunteers), be reintroduced on seed potatoes or tomato transplants, or live spores must blow in with rainstorms.

Disease development is favoured by cool, moist weather. Nights in the 50's and days in the 70's accompanied by rain, fog or heavy dew are ideal. Under these conditions, lesions may appear on leaves within 3-5 days of infection, followed by the white mold growth soon thereafter. Spores formed on the mold are spread readily by irrigation, rain and equipment. They are easily dislodged by wind and rain and can be blown into neighbouring fields within 5-10 miles or more, thus beginning another cycle of disease.

Infection of potato tubers arises from spores that develop on foliage. Tubers exposed by soil cracking or erosion of hills may come in contact with spores washed down from infected leaves and stems by rainfall or irrigation. Tubers infected during the growing season may partially decay before harvest. Tuber infection may also occur at harvest when tubers contact living spores remaining on infected vines. Little if any tuber-to-tuber spread of late blight occurs during storage if tubers are kept under cool, well-ventilated conditions.

The protectant fungicides commonly used to protect plants from late blight remain fully effective with all known strains of the fungus.

1. Infected cull potatoes are a major source of spores of the late blight fungus and must be disposed of properly-Do Not Make Cull Piles. Cull potatoes should be spread on fields not intended for potato production the following year in time that they will totally freeze and be destroyed during the winter. If this is not possible, they must be destroyed in some other way such as by complete chopping, burial, burning or feeding to livestock.
2. Plant only certified seed potatoes. Use of "year-out" seed or seed saved from local crops is asking for trouble with late blight. Seed sources should be selected very carefully to avoid bringing in late blight on seed potatoes, especially new strains of the fungus. Look for the characteristic coppery-brown discolouration of the potato flesh under the skin of seed tubers. Infected tomato transplants also can be a significant source of the disease. Use only obviously healthy tomato transplants free of dark lesions on leaves or stems.
3. Volunteer potatoes and tomatoes can be a significant source of spores of the late blight fungus. All volunteers should be destroyed as quickly as possible by herbicides, chopping, or cultivation.
4. Growers should scout fields regularly to look for late blight. Special attention should be paid to early-planted fields because that is where the disease is likely to develop first. Scouting should be concentrated in low-lying areas, field edges along creeks or ponds, near the centre of centre-pivot irrigation rigs, in areas near woodlots or any area that is protected from wind where the leaves tend to remain wet longer. Any area where it is difficult to apply fungicides such as edges and corners or under power lines if using aerial application should be examined. Scouts should look for large, black or purplish lesions on stems or leaves and the telltale

cottony, white mold growth, usually on the undersides. Be sure to check leaves and stems under the crop canopy as that is where the disease is most likely to begin.

5. Use of a good protectant fungicide programme is necessary to fully protect any crop of potatoes or tomatoes.
6. With potatoes, make sure that vines have been completely dead for 2-3 weeks prior to harvest. Fungicide applications should be continued until vines are dead. When foliage dies, spores of the late blight fungus that remain on the foliage also die. This practice will prevent infection of tubers during harvest and development of late blight in storage.

BACTERIAL RING ROT OF POTATOES

Bacterial ring rot is an important disease of potatoes and is one of the main reasons for rejection of seed potatoes from certification programmes. This disease is particularly serious because it has the potential to spread quickly throughout a farm and may lead to severe losses if left unchecked. Ring rot was originally found in Germany in the late 1800's. The causal bacteria were introduced into the United States in the early 1930's and by 1940 were found throughout the country.

Symptoms

Severe ring rot can result in wilting of leaves and stems along with yellowing and death of leaves. Lower leaves usually wilt first, are slightly rolled at the margins, and are paler green than healthy leaves. As wilting progresses, leaf tissues between veins become yellow. In the later stages of disease, margins of lower leaves die and become brittle, and eventually entire stems yellow and die.

Frequently, only one or two stems in a hill will develop symptoms and, in some cases, there are no above-ground symptoms at all. Ring rot derives its name from a characteristic breakdown of the vascular ring within the tuber. This often appears as a creamy-yellow to light-brown, cheesy rot. The symptom is most frequently observed when a diseased tuber is cut crosswise at the stem end.

In severe cases, the vascular ring may be separated, and a creamy or cheesy exudate can be forced out from this tissue when the tuber is squeezed. On the outer surface, severely diseased tubers may show slightly sunken, dry, cracked areas. Infected tubers are often invaded by secondary decay organisms which may lead to complete breakdown. Symptoms of ring rot in the vascular tissue of infected tubers are often less obvious t, appearing as only a broken, sporadically appearing dark line, or as a continuous, yellowish discolouration. Because of this, laboratory tests should always be performed to confirm a diagnosis of ring rot.

Causal Organism

Ring rot is caused by the bacterium Clavibacter michiganense subsp. sepedonicus. Ring-rot bacteria survive between seasons mainly in infected seed tubers. They are also capable of surviving 2-5 years in dried slime on surfaces of crates, bins, burlap sacks, or harvesting and grading machinery, even if exposed to temperatures well below freezing. Survival is longest under cool, dry conditions. Ring-rot bacteria do not survive in soil in the absence of potato debris, but can survive from season to season in volunteer potato plants. Wounds are necessary for penetration of the bacteria into seed pieces.

The pathogen is easily transmitted from diseased tubers to healthy seed pieces during the seed-cutting process. A knife that cuts one infected tuber can spread these bacteria to the next 20-100 seed pieces. Likewise, the bacteria may be spread during planting, particularly if a picker-type planter is used. Ring-rot bacteria can be moved in irrigation water and by chewing insects, such as Colorado potato beetles and flea beetles. After the bacteria become established in a plant, they multiply and move throughout the plant via the water-conducting tissues. Fortunately ring-rot bacteria are capable of causing disease only in potato, although they may be able to colonize roots of sugar beets.

1. Plant only certified disease-free seed tubers. Certified seed potatoes are produced under regulations mandating zero tolerance for ring rot. Although use of certified seed tubers will not guarantee total freedom from ring rot bacteria, it is the best assurance.
2. Discontinue use of any lot of seed tubers in which ring rot is found. Seed lots known to be contaminated with ring-rot bacteria should never be planted.
3. Before handling seed tubers, all containers, tools, knives and mechanical cutters, planters, and other equipment should be thoroughly washed with a detergent solution, rinsed, and then sanitized with a disinfectant (for current recommendations,
4. If ring rot is confirmed to be present, a thorough cleanup must be undertaken. Dispose of all infected tubers away from potato production areas. Clean all surfaces of storages and equipment to remove all mud, dirt and debris and then wash with a strong detergent in hot water applied by a high-pressure washer. After cleaning, sanitize all storages and equipment with a disinfectant. Do not plant potatoes for two seasons in any field in which ring rot has been found.

POTATO PINK ROT, PYTHIUM LEAK AND SEED-PIECE DECAY

Pink rot and Pythium leak, sometimes collectively called water rot, occur sporadically wherever potatoes are grown. These diseases are a problem of mature tubers at harvest and in storage. They are most serious when warm, wet soil conditions persist during tuber formation and at harvest. When newly-planted seed pieces are exposed to these conditions, Pythium seed-piece decay also can be severe. Major problems with these diseases are usually associated with excessive rainfall or irrigation either early or late in the season, especially on poorly-drained soils.

Symptoms

Pythium seed-piece decay often results in delayed emergence and poor stands. Infected seed pieces become a soft, watery mass in the soil. Symptoms of pink rot in mature plants include brown or blackened roots or stolons, and in severe cases, leaf chlorosis, stunting, wilting, and even plant death. Tubers develop pink rot mostly through diseased stolons, but occasionally infections occur at buds or lenticels. Decay spreads through infected tubers with the advancing margin of the rot usually sharply defined by a dark line, which may be visible through the skin. Eyes of infected tubers are often dark brown. Decaying tubers remain intact, but are spongy and odorless. If squeezed, a clear liquid will exude. When infected tubers are cut open, the internal tissues turn salmon pink after a 15-20 minute exposure to air, then later become brownish-black. Pythium leak usually develops through harvest wounds in tuber surfaces and begins as a discoloured, watersoaked area. As with pink rot, the advancing margin of infection is usually bounded by a dark line. Infected tissues are extremely watery, and appear brown or gray. Severely rotted tubers are of a uniform texture resembling a soft, watery paste.

Causal Organisms

Pink rot is caused by several species of the soilborne fungus Phytophthora (NOT the species that causes late blight) while Pythium leak and seed-piece decay are caused by several species of the closely related soilborne fungus Pythium. These fungi are widely distributed in both water and soil, and their behaviour is similar.

They survive in soil within decaying plant material, or as resistant spores free in the soil. In warm, moist soil, these fungi produce swimming spores that move in water films. Roots can be infected by Phytophthora at almost any stage of plant growth, but symptoms are more severe on younger roots. Both groups of fungi infect tubers through wounds, but Phytophthora generally infects tubers before harvest, often through stolons. Infection by Pythium usually occurs through harvest wounds, especially at temperatures above 70 F. Seed pieces can be infected by Pythium as soon as they are planted.

1. Select areas with well-drained soils for planting potatoes.
2. Use a crop rotation away from potatoes for at least 4 years if pink rot or leak have been severe. This may reduce the amount of fungus surviving in the soil.
3. Delay planting for at least 2 weeks after plowing down green vegetation as this may temporarily stimulate populations of Pythium fungi.
4. Avoid planting in soils colder than 45 F or warmer than 70 F.
5. Avoid harvesting infested fields when soils are especially wet or soil temperatures are below 50 F or above 65 F. Stop irrigation well in advance of harvest.
6. Avoid bruising tubers during harvest by adjusting equipment properly, keeping digger chains fully loaded and minimizing drops to 6 inches or less. Do not leave harvested tubers lying on warm, moist soils for any length of time as infection with Pythium may occur quickly.
7. Leave low spots in fields unharvested if they have been waterlogged and much rot is present.
8. Keep tubers cool and as dry as possible during harvest, loading, transit and storage.
9. Grade out infected tubers as much as possible before placing harvested tubers in storage.
10. Store lots of harvested tubers containing many infected tubers separately from healthy lots. Good

airflow through the pile should be provided to dry out leaky tubers. Lots with significant amounts of disease should be marketed as soon as possible as they will not store well.

SCAB OF POTATO TUBERS

Scab is a disease of potato tubers that results in lowered tuber quality due to scab-like surface lesions. There are no above-ground symptoms. Two forms of scab occur. Common scab occurs in all production areas and is most severe in soils with a pH above 5.5. Another less common form, called acid scab, is important in acidic soils.

Symptoms

Scab symptoms are quite variable. Usually, roughly circular, raised, tan to brown, corky lesions of varying size develop randomly across tuber surfaces. Sometimes scab develops as a rather superficial layer of corky tissues covering large areas of the tuber surface. This is called russet scab. Pitted scab can also occur where lesions develop up to 1/2 inch deep.

These deep lesions are dark brown to black, and the tissues underneath are often straw-coloured and somewhat translucent. More than one of these lesion types may be present on a single tuber.

Although scab symptoms are usually noticed late in the growing season or at harvest, tubers are susceptible to infection as soon as they are formed. Small brown, water-soaked, circular lesions are visible on tubers within a few weeks after infection. Mature tubers with a well-developed skin are no longer susceptible, but existing lesions will continue to expand as tubers enlarge. Thus disease severity increases throughout the growing season. Scab is most severe when tubers develop under warm, dry soil conditions. Coarse-textured soils that dry out quickly are therefore more conducive to scab than are fine-textured soils.

A few other conditions can be confused with scab. White, enlarged lenticles, which frequently occur on potato tubers harvested from wet soil, can be mistaken for scab. Usually this condition will disappear when tubers are dried. Patchy russeting, checking, or cracking of tuber surfaces caused by the fungus Rhizoctonia also may be confused with russet scab. A totally different but uncommon disease called powdery scab, caused by the fungus Spongospora subterranea, causes very similar scab-like symptoms. Laboratory examination may be necessary to identify these diseases.

Causal Organisms

Scab is caused by a group of filamentous bacteria called actinomycetes that occur commonly in soil. In soils with a pH above 5.5, Streptomyces scabies is usually responsible for common scab, and is capable of causing all the types of scab lesions. It is commonly introduced into fields on seed potatoes, and will survive indefinitely on decaying plant debris once the soil is contaminated. Because the organism can survive passage through the digestive tract of animals and be distributed.

Blackleg, Aerial Stem Rot, and Tuber Soft Rot of Potato

Blackleg, aerial stem rot, and tuber soft rot are all similar diseases caused by several types of soft-rot bacteria. Blackleg and tuber soft rot occur wherever potatoes are grown. Aerial stem rot is also widespread, but is most severe under sprinkler-irrigation.

Symptoms

Blackleg begins from a contaminated seed piece, but the symptoms can occur at several stages of plant development. In severe cases, entire seed pieces and developing sprouts may rot in the ground prior to emergence, resulting in a poor stand. Blackleg often develops after plants are well up or even in flower.

In this case, stem bases of diseased plants typically show an inky-black to light-brown decay that originates from the seed piece and can extend up the stem from less than an inch to more than two feet. Leaves of infected plants tend to roll upward at the margins, become yellow, wilt, and often die. Aerial stem rot (also called bacterial stem rot or aerial blackleg) is initiated by soft-rot bacteria from sources external to the seed piece. Stem infection can occur through wounds or through natural openings such as leaf scars. Lesions on diseased stems first appear as irregular brownish to inky-black areas.

These enlarge into a soft, mushy rot that causes entire stems to wilt and die. Potato tubers with soft rot have tissues that are very soft and watery, and have a slightly granular consistency. The diseased tissue is cream- to tan-coloured, and often has a black border separating diseased from healthy areas. In the early stages, soft-rot decay is generally odorless, but later a foul odor and a stringy or slimy decay usually develops as secondary decay bacteria invade infected tissues. Most internal tuber tissues may be consumed by soft rot organisms, sometimes leaving only a shell of skin remaining in the soil.

Causal Organisms

Blackleg, aerial stem rot, and tuber soft rot are caused by two closely related bacteria, Erwinia carotovora subsp. atroseptica and Erwinia carotovora subsp. carotovora. E. c. carotovora is very common and has an extensive host range, including most fleshy vegetables. It survives readily in soil and surface waters such as rivers, lakes, and even oceans. These bacteria are capable of multiplying and persisting in the root zones of many host and nonhost crop and weed species. In contrast, E. c. atroseptica is associated mostly with potatoes.

These bacteria do not survive well in soil for more than one year, unless they are contained within diseased tubers or other potato plant debris. Blackleg is usually caused by E. c. atroseptica carried on contaminated seed tubers. Most lots of seed tubers are contaminated to some degree, but the bacteria are usually dormant and do not cause disease unless environmental conditions are favourable. In contrast, aerial stem rot is usually caused by E. c. carotovora contained in infested soil or introduced to the crop by irrigation water, wind-blown rain, and insects. Tuber soft rot can be caused by either of these soft-rot bacteria.

Moisture and temperature are the two critical factors in initiation and development of soft-rot diseases. High soil temperatures and bruising of seed tubers favour seed-piece decay and pre-emergence blackleg. Blackleg in growing plants is favoured by cool, wet soils at planting followed by high temperatures after emergence.

Dense plant canopies and long periods of leaf wetness favour infection of aerial plant parts. Although tuber soft rot can occur at any temperature above 50 F, disease develops best above 75 F. Oxygen depletion in tubers also favours soft rot. When seed pieces in soil or tubers in storage become covered with a film of water, the tissues rapidly become depleted of oxygen. This also may be induced by soil flooding or improper drying of washed tubers. Once it starts, tuber soft rot can proceed rapidly in storage. "Wet" areas may develop in the piled tubers that flow onto ones below, spreading the bacteria. Heat, coupled with condensation on tuber surfaces, can further adversely affect storage conditions, resulting in accelerated "melt" of the pile.

1. Plant only certified, disease-free seed tubers. If possible, use whole (B-size) seed tubers that do not have to be cut.
2. When receiving seed tubers in bags, do not stack more than five bags high. With bulk or bagged seed, store at 40-45 F until 2-3 weeks before planting, then warm to 55-60 prior to cutting.
3. Clean all equipment used for cutting seed tubers thoroughly and then sanitize with an appropriate disinfectant.
4. Treat cut seed pieces with recommended fungicide dressings immediately after cutting.
5. Plant treated cut seed pieces immediately if soil temperatures are 55-65 F at planting depth. Seed

pieces can be held 1-2 weeks at 55-60 F and 95-99 per cent relative humidity to hasten healing of cut surfaces. Condensation on surfaces of seed pieces must be avoided.

6. Do not irrigate fields until plants are well emerged. Avoid using surface water for irrigation.
7. During crop growth, monitor irrigation and nitrogen fertility to minimize excessive vine growth that will promote leaf wetness within the plant canopy.
8. Harvest tubers only after the vines are completely dead to ensure skin maturity. Low spots in the field should be left unharvested if significant waterlogging has occurred.
9. Take all precautions to minimize cuts and bruises when harvesting and handling tubers.
10. Hold newly harvested potatoes at 55-60 F with 90-95 per cent relative humidity for the first 1-2 weeks to promote wound healing. After this curing period, lower the temperature of table stock to 38-40 F for long-term storage. Never wash tubers prior to storage.

FUSARIUM DRY ROT AND SEED-PIECE DECAY OF POTATO

Fusarium dry rot is an important postharvest disease of potato tubers that causes significant losses in storage and transit of both seed tubers and those for table consumption. It is also a major cause of seed-piece decay after planting.

Symptoms

Infected tubers usually develop a dry rot, but a moist rot may occur if secondary infections with soft-rot bacteria also are involved. Surfaces of infected tubers are sunken or wrinkled, and rotted tissues appear brown or gray to black. A white or pink mold is sometimes visible on tuber surfaces. When tubers are cut, internal cavities within rotted tissues may contain white, yellow or pink molds. In storage, blue, black, purple, gray, white, yellow, or pink spore masses may develop in these internal cavities. After low-temperature storage, internal tissues often will become firm and dry or even powdery.

Causal Organisms

Fusarium dry rot is caused by several species of the soilborne fungus Fusarium. These fungi are common in most soils where potatoes are grown and survive as resistant spores free in the soil or within decayed plant tissues. Although some infections may develop on tubers before harvest, most infections occur as the fungus enters tubers through harvest wounds. Small, brown lesions appear at wound sites 3-4 weeks after harvest and continue to enlarge during storage, taking several months to develop fully. The disease develops fairly rapidly at temperatures above 50 F, but lesions will cease enlarging below 40 F. The fungus is only dormant at these low temperatures, however, and will resume growth when tubers are warmed.

Fusarium seed-piece decay is really the same disease. Seed tubers may be infected prior to shipment. Decay during transit or storage often accounts for poor quality seed tubers. When these Fusarium fungi are present on seed pieces or in the soil, poor stands may result, especially if cut surfaces of seed pieces are not properly healed. Fusarium seed-piece decay begins as reddish-brown to black depressions on cut surfaces. These may expand to cover the entire seed piece and often result in a slimy rot when infection by secondary soft-rot bacteria follows.

1. Harvest tubers only after the vines are completely dead to ensure skin maturity.
2. Take all precautions when harvesting and handling tubers to minimize cuts and bruises.
3. Hold newly harvested potatoes at 55-60 F with 90-95 per cent relative humidity for the first 1-2 weeks to promote wound healing. After this curing period, lower the temperature of table stock to 38-40 F for long-term storage.

4. Plant only certified, disease-free seed tubers. If possible, use whole (B-size) seed tubers that do not have to be cut into seed pieces before planting.
5. When receiving seed tubers in bags, do not stack more than five bags high. With bulk or bagged seed, store at 40-45 F until 2-3 weeks before planting. Then allow seed potatoes to warm prior to cutting.
6. Treat cut seed pieces with recommended fungicide dressings immediately after cutting.
7. Plant treated cut seed pieces immediately or store them at 55-60 F and 95-99 per cent relative humidity to hasten healing of cut surfaces. Condensation on surfaces of seed pieces must be avoided.

RHIZOCTONIA STEM AND STOLON CANKER OF POTATO

Rhizoctonia stem and stolon canker, also called black scurf, is an extremely common problem on potatoes that can result in delayed emergence, reduced stands and poor tuber quality.

Symptoms

Rhizoctonia canker often goes unnoticed until harvest when tubers are found to be covered with small brownish-black fungal bodies (sclerotia) that look somewhat like bits of black soil that will not wash off. These may vary in size from a pinhead to as large as a pea. When introduced into soil, these structures can survive and germinate the following spring to attack young shoots, roots, stolons, and tubers of the new crop. In many cases, weak, spindly-looking or late-emerging plants may be the result of an attack by this fungus.

The first sprouts are often killed before they reach the surface, resulting in the emergence of a weaker secondary sprout. Dry, sunken, brownish lesions developing on the base of the stem below the soil line are evidence of the stem canker phase of this disease.

The lesions may girdle the stem or large cankers may interfere with movement of nutrients from the leaves to the tubers. In such cases, vines become yellow to reddish purple; the leaves begin to curl upward; the stalks swell, particularly at the nodes; and often small purplish tubers form where leaves branch from stems. During midseason under a dense canopy of foliage, the fungus may develop a white, powdery mold growth on stems, extending just above the soil line. Cankers that form on stolons may prune off young developing tubers. This disease may lead to russeting or surface cracking of mature tubers and sometimes shallow, brown lesions will form around lenticles. Low soil temperatures in the fall favour formation of sclerotia on tubers.

Causal Organism

This disease is caused by the fungus Rhizoctonia solani which can survive in the soil for many years, even under relatively dry conditions. Rhizoctonia causes disease in a wide variety of crops, but the strains found in association with potato generally do not attack and reproduce on other plant species. The fungus survives in soil associated with decomposing plant residues. In addition, the sclerotia can survive on infected tubers and can persist free in soil for extended periods.

Emerging sprouts are usually attacked by fungus present on the seed tubers. Once green leaves develop on sprouts, stem tissues are much less susceptible to infection. Stem cankers, stolon infections, and sclerotia on tubers usually develop when these tissues grow in proximity to sources of the Rhizoctonia fungus in soil. Cool (55-60 F), moist soils are optimal for infection. Sclerotia form on the surfaces of mature tubers under cool, moist conditions, generally after the vines have begun to die.

1. Use a crop rotation with corn, grasses, and cereal grains. If this disease has been severe, 3-5 years should elapse between potato crops.

2. Plant certified seed tubers that are free of Rhizoctonia on the skin.
3. Use planting practices that promote rapid emergence: Avoid planting in heavy, poorly-drained soils. Plant seed potatoes when soil is warm (above 60 F). Cover seed tubers with no more than 2 inches of soil.
4. Harvest tubers promptly after vines are dead to avoid the development of sclerotia on the surfaces of tubers while still in the soil.

MOSAIC VIRUS DISEASES OF VINE CROPS

Mosaic diseases of vine crops are caused by at least five different viruses: 1) cucumber mosaic virus (CMV), 2) squash mosaic virus (SQMV), 3) watermelon mosaic virus-2 (WMV-2), 4) zucchini yellow mosaic virus (ZYMV) and 5) papaya ringspot virus-watermelon isolate (PRSV-W). All five of these viruses can infect all cultivated vine crops (squash, melons, gourds, cucumbers, pumpkins) and under ideal conditions can cause a high rate of crop failure and severe economic losses.

Symptoms

As the names imply, all five viruses cause a mottling of foliage called mosaic. This is characterized by the presence of intermingled patches of normal and light green or yellowish coloured plant tissue. Depending on environmental conditions, mosaic symptoms can range from mild to severe and be visible on both leaves and fruit. The younger the plant when infected, the more severe the symptoms as the plant matures. In some cases, plants infected at the seedling stage may collapse and die. Plants infected at the flowering stage may not set fruit or young fruits may abort.

If plants are more mature when infected they do not show severe mosaic and may still produce marketable fruit. The most dramatic symptoms are often associated with infected fruit. Fruit symptoms can range from subtle colour change to severe deformation. It is not uncommon to have two or more viruses infecting the same plant and in these cases symptoms may be much more severe than if the plant were infected with only one virus. It is nearly impossible to distinguish between any of the five viruses based only on visible symptoms. Because of this, it is important to have suspect plants sent to a diagnostic clinic capable of using specialized techniques to identify the viruses that are present. Proper disease management depends on knowing which specific virus is involved.

Disease Cycle and Transmission

All five vine crop mosaic viruses are transmitted by insects. CMV, ZYMV, WMV-2 and PRSV-W are transmitted primarily by the green peach aphid and the melon aphid. SQMV is transmitted by the striped cucumber beetle and spotted cucumber beetle and also through seed, but only at a very low rate. In most cases these insects acquire the virus by feeding on weeds or other vine crops that are infected. The insects then feed on noninfected vine crops and in the process of this feeding transmit the virus to the host plant. In the case of the green peach and melon aphid, the insect carries the virus on its probing mouthparts and the aphid only needs to probe the plant to transmit the virus.

Feeding by the aphid is not necessary. Once the insect probes one plant, it can move onto the next, probing and infecting as it goes. The more aphids present, the quicker and more complete the virus spread. By this process, large acreages can become infected quickly.

1. Plant varieties with resistance to these viruses whenever possible. The availability of varieties with virus resistance varies depending on the virus and the vine crop in question. Major seed producers list resistant varieties in their seed catalogs.
2. Control weeds which could harbor the viruses from areas around production fields. Pokeweed is a primary weed in the area which harbors many plant

viruses. Weeds also serve as hosts for insects that transmit the viruses.

3. Since the virus is moved about by insects it would seem logical that insect control would be a primary method of virus control. However, in this case it is not that simple. Since aphid vectors need only probe the plant to transmit the virus, insect control is not really that effective in controlling the virus. Even if the aphid were to die immediately following probing, virus transmission would still take place. To control the insect effectively insecticides would need to be applied on a daily basis. The economics of this may not be feasible. Aphids also move long distances with weather fronts making local control difficult.

DAMPING-OFF AND ROOT ROT OF BEANS

Root rots, damping-off before and after seedling emergence, and seed rots are destructive diseases of green, snap, lima, and dry beans. These diseases are caused primarily by soilborne fungi.

Significant losses may occur to susceptible varieties, especially if cool, wet weather conditions prevail for the first few weeks after seeding and then are followed by hot, dry weather. Disease incidence and severity often vary greatly, even in areas with a history of root rot. In the same growing season, it is not uncommon to lose a crop completely and then re-seed and experience no problems. This situation results from changes in biological, environmental, and soil conditions. Since there are no commercially acceptable resistant varieties, growers should learn how to recognize these diseases and use a combination of management practices to minimize potential losses.

Symptoms

Damping-off before emergence results from fungal attack of germinating seeds and/or young seedlings while they are still in the ground. Infected seeds may fail to germinate, become soft and mushy, and finally disintegrate. Slightly darkened water-soaked lesions may be visible on stems of young seedlings. Infected areas enlarge quite rapidly, and seedlings may die shortly after infection, prior to emergence from soil. Roots or stems of seedlings that have already emerged also can be attacked at or below the soil line resulting in damping-off.

Infected roots are usually discoloured or rotted and sometimes reddish-brown lesions develop on the tap root. Infected stem tissues are soft and colourless to dark-brown. Basal portions of invaded stems may be much thinner than the areas above the lesion, a condition called "wire stem." As a result, the seedling may fall over and die. Damping-off is a major cause of poor stand establishment in bean plantings. Older plants can also be attacked by these fungi. Later infections are usually confined to roots, which may result in stunting, wilting, or plant death. To diagnose bean root rots, suspected plants should be carefully dug and washed, because pulling plants may leave tissues with characteristic symptoms in the soil. If plants are brought to a diagnostic clinic, they should be dug and left intact in soil.

Causal Organisms

Root rots, damping-off, and seed rots are caused primarily by soilborne fungi. Fusarium root rot is caused by Fusarium solani f. sp. phaseoli. This fungus is capable of surviving long periods in soil, even in the absence of beans, by the production of thick-walled resting spores. Its host is mainly green beans, but lima beans and garden peas are also susceptible. Coarse-textured, acidic and poorly fertilized soils favour development of Fusarium root rot. A number of fungi in the genus Pythium are capable of inciting seed rot and seedling damping-off. This may vary with environmental conditions and developmental stage of the host.

These fungi can survive in soil for many years, either by producing thick-walled resting spores, or in a vegetative condition within crop residues left from previous years. They may also survive by attacking several other garden vegetable crops such as beets, cabbage, peas, melons, squash, and cucumbers.

Development of diseases caused by Pythium is more prominent under:

1. High or low temperatures that are unfavourable to crop growth,
2. Excess water and/or nitrogen,
3. Continued cropping of susceptible plants.

Rhizoctonia solani is another soil-inhabiting fungus causing root rots and damping-off. This fungus overwinters free in the soil or within decayed plant tissues. Aside from bean, its host range includes beets, cabbage, lettuce, peas, pepper, tomatoes, and many others. Warm, relatively dry soil conditions favour disease development.

1. Do not grow beans or other susceptible crops continually in the same location. Continuous cropping of susceptible plants will eventually lead to a buildup of these fungi in the soil. Since they are capable of long-term survival, a rotation of 4-5 years is desirable. Avoid planting beans in fields known to be heavily infested with bean root-rot fungi.
2. Plant beans only on well-drained soils or try to improve drainage. This could be done by improving soil structure and/or installing drain tiles. Subsoiling to a depth below the plowed layer will reduce soil compaction, and improve drainage. Deep plowing of the previous years' crop residues will reduce bean root rot.
3. Delay planting until the soil is warm (above 65 F) and seed shallow to insure rapid emergence. Avoid planting seeds too close together-follow instructions on the seed container. Do not overfertilize, especially with nitrogen.
4. Use of fungicide-treated seeds will minimize problems with damping off and root rots.

POWDERY MILDEW OF VINE CROPS

Powdery mildew can be a serious problem on foliage and stems of cucumbers, melons, squash, pumpkins, and ornamental gourds.

Symptoms

Small patches of fine, white threads develop on surfaces of infected leaf blades in late July on most vine crops. These patches grow together and eventually cover stems and foliage with white, powdery masses of spores. In severe infections, leaves will yellow and die. Petioles, stems and, rarely, fruit will also become infected. Early death of leaves can decrease the total amount of fruits formed as well as reduce fruit size. Loss of foliage hastens maturity of fruit and increases sunburning. Stems infected with Powdery Mildew become brown. This can spoil the appearance of the "handles" on pumpkins which lowers value of the fruit.

Causal Organisms

Two fungi cause powdery mildew on vine crops, Sphaerotheca fuliginea and Erysiphe cichoracearum. S. fuliginea is most common. Powdery mildew fungi thrive under conditions of high relative humidity, warm temperatures, low light, high fertility, and succulent plant growth. Unlike bacteria and many fungi, free moisture on leaf surfaces actually inhibits infection by these pathogens, although very high relative humidity is required for spore germination. Powdery mildew fungi grow only on living host plants. They survive the winter as dormant mycelium on perennial plants or as spores in thick-walled fruiting structures. Powdery Mildew spores can be blown into the spring from warmer southern areas. Mature foliage is most readily infected; very young leaves are nearly immune.

Control

1. Plant resistant cultivars of cucumbers and muskmelons. Pumpkin varieties vary greatly in susceptibility to this disease.

2. Use cultural practices that avoid excessive succulence, overcrowding, shading, overwatering, or excess fertilization especially with nitrogen.
3. Avoid making new planting of vine crops in the vicinity of older plantings, especially if mildew is already present.
4. Beginning in late July or early August when Powdery Mildew first appears, regular applications of fungicides may be required.

SEPTORIA LEAF SPOT OF TOMATOES

Septoria leaf spot is a disease of the foliage and stems. It does not affect fruit directly. The disease causes rapid defoliation when weather is warm and moist.

Symptoms

Although the symptoms may appear on the leaves and stems at any stage of plant development, they usually become evident after plants have begun to set fruit in mid-July. Initially the fungus causes small, water-soaked, roughly circular spots scattered randomly over the leaf. These spots enlarge to become l/16 to l/8 inch in diameter with dark margins and tan centres. Small, dark, pimple-like fruiting bodies in which spores of the fungus are produced can be seen in the centre of the lesions. Older leaves near the ground usually show symptoms first. Symptoms appear rapidly on younger foliage in rainy weather. When leaves are heavily infected, they drop prematurely. Exposed fruits are subject to sunscald. The fungus which causes this disease can grow on weeds such as Jimson weed, horse nettle and nightshade.

Causal Organism

Septoria lycopersici lives between tomato crops in the soil on infested debris of tomato and weeds. Spores formed on crop debris splash onto foliage and start the disease. Wind and rain spread spores produced in the dark bodies formed in leaf spots to adjacent uninfected leaves. The fungus is most active between 60 and 80 degrees F when rainfall is abundant.

Control

1. Rotate out of tomatoes for 4 years.
2. Deep plow, preferably in the fall, to bury all plant refuse.
3. Grow tomato transplants in sterilized soil.
4. Control weeds, especially horse nettle, Jimson weed, and nightshade.
5. Use of a protectant fungicide may be necessary to adequately control Septoria Leaf Spot when conditions are favourable for disease development.

FUSARIUM WILT OF VINE CROPS

Several different species of Fusarium, a soil-borne fungus, cause wilting of watermelon, muskmelon, cucumber, squash, and other vine crops. In many cases the fungus-causing wilt in a particular crop is specific to that crop. These fungi are generally capable of surviving for long periods in the soil.

Symptoms

Wilting of older plants is often the first symptom of the disease. Before diseased plants totally collapse, however, they may begin to wilt during the hottest part of the day and recover during the night. Infected plants are often stunted and yellowed. Leaves often have dead areas which can mimic nutrient deficiencies. Stems of wilted plants when cut lengthwise at the soil line may show brown discolouration in the woody tissues immediately under the bark. Vines killed by Fusarium can be covered with pinkish-white fungal growth in wet weather.

Causal Organism

Fusarium oxysporum is well adapted to life in the soil. It can survive season to season in old diseased vines or it can live free in infested soil for many years in the absence of its host crop. The fungus grows at soil moisture and temperature favourable for vine crop growth. If the soil is very wet, infection is reduced. The fungus is stimulated to germinate when roots of susceptible host plants are growing nearby. It enters the plant through root tips or where some opening is present.

Control

1. Plant wilt-resistant cultivars whenever possible. Degree of resistance is influenced by the populations of the fungi in the soil, and which races are present.
2. Avoid introduction of the fungus into fields. Once soil is infested, crop rotation will be of little use because of the long-term survival of these fungi in soil. The fungus can be spread on equipment, tools, feet, and surface water contaminated with infested soil. Do not put compost on fields which has been made from diseased vines. Compost resulting from such piles will contain the fungus.
3. Crop rotations of three to four years may be helpful in lowering the amount of Fusarium in the soil. Since the types of Fusarium-causing wilt in a crop are generally specific to that crop it is possible to rotate between vine crops in infested soil.
4. In commercial production of vine crops, soil fumigation may be useful.

ANTHRACNOSE OF TOMATO

Anthracnose is a common and serious disease of tomato fruit. This disease can occasionally cause severe damage to peppers, especially when red fruit is allowed to develop. Anthracnose can reduce a bountiful harvest into rotted fruit in a few days in warm, moist weather.

Symptoms

Small, watersoaked, circular lesions develop under the skin of fruit as it ripens. These become sunken and dark. Numerous dark specks, the fruiting bodies of the fungus, develop in the lesions in concentric rings. In moist, warm weather, these black bodies ooze gelatinous pink spore masses. In warm weather the fungus and soft rot bacteria which enter the split skin over the lesions spread internally forming a semisoft decay which renders the fruit worthless.

Causal Organism

Colletotrichum coccodes survives between crops on infested plant debris in the soil. Early in the growing season, spores from the soil splash on lower leaves of the tomato plant. Few symptoms develop on infected leaves, but the spores produced on foliage can be carried by splashing rain to developing green fruit. Infected green fruit will not develop symptoms of anthracnose until they begin to ripen. Ripe fruit is very susceptible to this fungus.

Control

1. Rotate 3 years between pepper and tomato crops.
2. Plant tomatoes and peppers in well drained fields to avoid excess soil moisture as fruit ripen.
3. Apply overhead irrigation during the early part of the day so that plants dry before sundown.
4. Harvest and use fruit before it fully ripens.
5. If conditions favour development of anthracnose, a preventative spray programme may be required to give adequate control of this disease. Apply registered fungicides according to product label instructions when weather conditions are above 65 degrees F and the foliage is likely to remain wet longer than 6 hours. Applications to tomatoes should begin when the first fruit is larger than a walnut. Applications to peppers should be started as soon as fruit is present.

ROOT KNOT NEMATODE

Root knot first attracted attention in 1855 when it was found in a greenhouse in England. Since then it has been a cause for concern over much of the world. It affects more than 2,000 species of plants including economically important forage crops, small grains, fruits, vegetables, field crops, nursery crops, and turf grasses. Root knot nematodes are cosmopolitan in distribution and occur widely in the United States.

Symptoms

Root knot is very distinctive because of the galls or swellings produced on roots and underground portions of stems. These deformations can often completely ruin crops for sale. Plants, if infected when young, will be stunted, more susceptible to drought stress, and show symptoms of nutrient deficiency.

Large and small roots may be affected with swellings varying from round sphere-like galls to elongated spindles formed from large numbers of individual galls growing together. Root knot galls involve the entire root in the affected zone. They do not take the form of easily detached galls like those produced by nitrogen-fixing bacteria on the roots of legumes. Root knot galls should not be confused with the fungal disease, club root, on cruciferous crops. In most instances, root knot is characterized by smaller swellings, and more uniformly distributed infection on the lateral feeding roots than is typically seen with clubroot.

When the galls formed by the root knot nematode are broken open, shiny white bodies about the size of a pinhead, the enlarged female nematodes, are usually found. Also, glistening white to yellow egg masses are present on the root surface. The galls formed by clubroot do not possess this characteristic and are usually pinkish or brick coloured. Phenoxy-type herbicides such as 2,4 D can cause swellings on stems of cruciferous crops which superficially look like root knot galls. Herbicide damage on these crops will not affect the roots, however. On young potato tubers the outer surface appears rough and warty because of the enlarged females underneath the skin.

Causal Organism

Root knot is caused by a small round worm, Meloidogyne spp. There are about 40 species at the present time. Not all plants are susceptible to any one species.

Juveniles emerging from eggs in the soil penetrate between and through cells to a position at the centre of the root usually near the growing tip. Feeding by the nematodes causes increases in root cell numbers and size. Enlarged cells, called giant cells, serve as food sources for female root knot nematodes.

During feeding, juveniles which will become females undergo a series of molts and enlarge. The female does not move from the feeding site. Each female deposits 300-500 eggs in a protective, jelly-like matrix at the root surface. Eggs, particularly in the egg mass, can withstand unfavourable environmental conditions. They represent one means of overwintering. Root knot nematodes may also overwinter in the soil in a juvenile stage. Root knot nematodes are often introduced into a growing area in infected planting stock.

Control

1. Rotate at least 3 years with non-host crops. While the feeding habits of field populations must be determined before crop rotation plans are set, the most common root knot species will not reproduce on grasses, small grains, or corn. Clean cultivation to keep down weed hosts of the nematodes is important.
2. Use transplants and nursery stock certified free of root knot nematode.
3. Plant resistant varieties wherever possible.
4. In commercial production, soil can be treated with fumigant-type nematicides.

PHYTOPHTHORA BLIGHT OF PEPPER AND CUCURBITS

Phytophthora blight, a highly destructive disease of peppers and cucurbits. It can become a serious problem during periods of heavy rainfall; the pathogen can spread rapidly through the crop, resulting in severe losses within a short time.

Symptoms

Peppers

Phytophthora blight affects both seedlings and mature plants. Infected seedlings show typical damping-off symptoms. Infection of older plants usually begins at or below the soil line. Water-soaked, dark brown lesions on the lower stems (collar rot phase) usually extend upward for an inch or more above the soil line and may expand to girdle the stems, preventing upward movement of water and nutrients.

This often results in a sudden wilting of foliage. Root infections may also occur which kill roots and cause wilting of the plant without the appearance of stem cankers. The foliar phase of this disease commonly occurs at forks in the branches, resulting in dark, girdling cankers and wilting of leaves and fruits. Infected leaves develop circular or irregular, dark green, water-soaked lesions which dry and appear light tan. Fruit lesions may also appear as enlarging, watersoaked areas, which then shrivel and darken. A mass of white fungal growth may develop inside the fruit, and seeds usually turn dark brown or black. A fine, grayish-white to tan mold may also become evident over the lesion on the fruit surface. Under humid conditions, fungal growth develops extensively over the entire fruit.

Cucurbits

Seedlings show typical symptoms of damping-off. Mature plants may wilt suddenly as a result of root infection, in the absence of obvious stem or vine lesions. Stem and leaf petiole lesions are light to dark brown, watersoaked and irregular, eventually becoming dried and papery. Fruit symptoms begin as small, water-soaked circular lesions, which expand to become large, soft, sunken areas covered with white fungal growth. Infected fruit often collapse or "melt down" in the field or in storage.

Causal Organisms

Phytophthora blight is caused principally by the soil-borne fungus, Phytophthora capsici, although a similar fungus, Phytophthora parasitica, has also been reported to cause fruit rot on peppers and cucurbits. Both of these fungi are dependent on free water in soil for infection. Because of this, initial infection by Phytophthora usually occurs on plants growing in poorly drained areas of fields.

The fungi that cause Phytophthora blight survive as thick-walled, resistant spores (oospores) in the soil and as mycelium in infected plant tissues. Once introduced, these fungi can survive up to 15 months in moist soil in the absence of host plants. They can also be carried on seed or transplants. Once stem infection occurs, the fungi produce spores on infected stem tissues, which are then carried by splashed rain onto nearby plants.

Lower branches of adjacent plants can also be infected by rain-splashed soil contaminated by run-off water. Fruits in contact with the soil are especially prone to infection. Spores are produced on newly infected fruit and stems, and new infections can develop quickly in a short time. Warm (68 to 75 degrees F), wet weather is most favourable for disease development and spread of the pathogen. Disease progress declines when dry weather returns.

1. Use only certified disease-free seed or transplants.
2. Produce pepper plants on raised beds to retard initial stem infection. Arrange beds to prevent puddle formation between rows. Avoid depressions on the surface of the beds where water can accumulate, especially around the base of plants. Avoid planting peppers or cucurbits in poorly drained fields.
3. Practice crop rotation so that peppers or cucurbits are grown only every 3 to 4 years to reduce the amount of Phytophthora present in the soil. Use non-susceptible crops such as corn, small grains, beans, crucifers, or potatoes in the rotation.

4. Application of fungicides may reduce disease development and spread if used in combination with cultural practices such as crop rotation and raised beds. Application at planting and two or more (depending on the fungicide used) applications after planting are required.

BLOSSOM-END ROT OF TOMATO, PEPPER, AND EGGPLANT

Blossom-end rot is a serious disorder of tomato, pepper, and eggplant. Growers often are distressed to notice that a dry sunken decay has developed on the blossom end of many fruit, especially the first fruit of the season. This nonparasitic disorder can be very damaging, with losses of 50 per cent or more in some years.

Symptoms

On tomato and eggplant, blossom-end rot usually begins as a small water-soaked area at the blossom end of the fruit. This may appear while the fruit is green or during ripening. As the lesion develops, it enlarges, becomes sunken and turns black and leathery. In severe cases, it may completely cover the lower half of the fruit, becoming flat or concave. Secondary pathogens commonly invade the lesion, often resulting in complete destruction of the infected fruit. On peppers, the affected area appears tan, and is sometimes mistaken for sunscald, which is white. Secondary molds often colonize the affected area, resulting in a dark brown or black appearance. Blossom end rot also occurs on the sides of the pepper fruit near the blossom end.

Cause

Blossom-end rot is not caused by a parasitic organism but is a physiologic disorder associated with a low concentration of calcium in the fruit. Calcium is required in relatively large concentrations for normal cell growth. When a rapidly growing fruit is deprived of necessary calcium, the tissues break down, leaving the characteristic dry, sunken lesion at the blossom end. Blossom-end rot is induced when demand for calcium exceeds supply. This may result from low calcium levels or high amounts of competitive cations in the soil, drought stress, or excessive soil moisture fluctuations which reduce uptake and movement of calcium into the plant, or rapid, vegetative growth due to excessive nitrogen fertilization.

1. Maintain the soil pH around 6.5. Liming will supply calcium and will increase the ratio of calcium ions to other competitive ions in the soil.
2. Use nitrate nitrogen as the fertilizer nitrogen source. Ammoniacal nitrogen may increase blossom-end rot as excess ammonium ions reduce calcium uptake. Avoid over-fertilization as side dressings during early fruiting, especially with ammoniacal forms of nitrogen.
3. Avoid drought stress and wide fluctuations in soil moisture by using mulches and/or irrigation. Plants generally need about one inch of moisture per week from rain or irrigation for proper growth and development.
4. Foliar applications of calcium, which are often advocated, are of little value because of poor absorption and movement to fruit where it is needed.

CLUBROOT OF CRUCIFERS

Clubroot is a world-wide problem in temperate climates in the production of cruciferous vegetables such as cabbage, broccoli, cauliflower, radishes, and turnips; and field crops such as mustard and rape. The disease was known as early as the 13th century in England where it was called "finger and toe" disease because of the shape of infected roots.

Symptoms

The most striking symptom of clubroot is an abnormal enlargement of the root system with clubs often thickest at the centre, tapering spindle-like towards the ends. In radishes, clubroot causes distorted swellings on the base of the bulb and along the tap root. In severe cases, entire plantings are destroyed. Clubroot-infected plants often wilt on sunny days and permanent wilting may accompany advanced decay of infected roots. Severe stunting may be evident if infection occurs early and the disease progresses rapidly. The malformed and greatly enlarged roots are the key symptom of this disease.

Causal Organism

Clubroot is caused by the soil-borne fungus, Plasmodiophora brassicae, which only infects plants in the crucifer family. It infects susceptible host plants through root hairs. Once in the tissue, it stimulates abnormal growth of affected parts, resulting in a swollen club. Infection is favoured by excess soil moisture and low pH, although it can occur over a wide range of conditions.

Once a plant is infected, numerous resistant spores of the fungus are produced in the "clubbed" tissues. As these tissues decay, spores are released into the soil where they can remain infectious for at least 10 years. Contaminated soil moved by wind or water can serve as a source of infestation of nearby fields causing outbreaks of disease in areas where susceptible crops are planted for the first time. Numerous races of the pathogen have been identified.

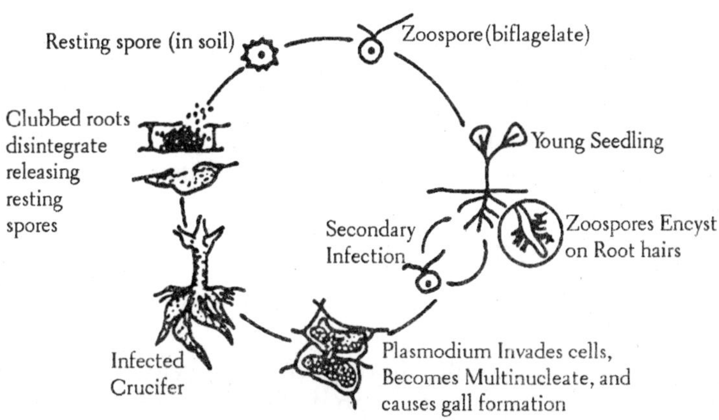

Clubroot is an very difficult disease to manage, and heavily infested areas may have to be abandoned for crucifer production. Some control may be achieved with the following measures:

1. A good crop rotation programme, growing crucifers on the same soil no more than every third or fourth year, is essential to retard development of a large population of spores on land not already heavily infested.
2. Since clubroot is favoured by a low pH, liming soil to pH 7.2 or above may be helpful. Raising soil pH too high, however, may interfere with the growth of succeeding crops other than crucifers. Calcitic lime is usually preferable to dolomitic lime, except for soils low in magnesium, where dolomitic lime is more effective. In course-textured soils, increasing the pH can result in boron deficiency. This may be alleviated by application of boron in transplant water or as a foliar spray.
3. With transplanted crops, the use of pathogen-free seedbeds and uninfected plants is essential to prevent introduction of the disease.
4. Application of a fungicide in transplant water or rototilled in a band prior to planting may help to reduce disease development.
5. Clean and disinfect all machinery before moving it from infested to non-infested land.
6. Some resistant cultivars are available. However, plant resistance has not been very useful in clubroot control because of rapid development of new races of the fungus.

COMMON SMUT OF CORN

Corn smut is an extremely common disease of sweet, pop, and dent corn and throughout the world. It is usually not economically important, although in some years yield losses in sweet corn may be as high as 20 per cent. In Mexico, immature smut galls are consumed as an edible delicacy known as cuitlacoche, and sweet corn smut galls have become a high value crop for some growers in the NE United States who sell them to Mexican restaurants.

Symptoms

The corn plant may be infected at any time in the early stages of growth, but becomes less susceptible after formation of the ear. Above-ground parts may be infected, but it is more common to see smut galls on the ears, tassels, and nodes than on the leaves, internodes, and aerial roots. The smut gall is composed of a great mass of black, greasy, or powdery spores enclosed by a smooth white covering of corn tissue. The gall may be 4-5 inches in diameter. When leaves are infected, small pustules develop, usually on the midrib, causing some leaf distortion. After the spores mature, the outer covering becomes dry and brittle, breaks open, and the spores sift out. Greatest yield losses occur when the ear becomes infected or if smut galls form on the stalks immediately above the ears.

Causal Organism

Corn smut is caused by the fungus, Ustilago zeae, that survives as a resistant spore in the soil over winter, and possibly for 2 to 3 years. These spores, called teliospores, can be blown long distances with soil particles or carried into a new area on unshelled seed corn and in manure from animals that are fed infected corn stalks. The teliospores germinate in moist air and give rise to tiny spores called sporidia. The sporidia bud like yeast, forming new spores that germinate in rain water that collects in the leaf sheaths. This leads to infections that are visible in 10 days or more. Wounds from various injuries provide points for the fungus to enter the plant.

The smut fungus is sensitive to temperature and moisture changes. In a warm season, the amount of smut is related closely to the amount of soil moisture, especially during June. When temperatures are lower than normal, there may be little smut even though soil moisture remains high.

1. Seed treatment is of no value because few spores are on the corn seed.
2. For the home garden, remove smut galls before they break open and bury or burn them. This must be done on a community basis, however, in order to be effective. Smut gall removal is not practical in commercial production.
3. In order to reduce infection points from insect injury, control corn borers as first tassels appear by application of insecticides when insect populations are high.
4. Avoid injury of roots, stalks, and leaves during cultivation.
5. Deep plow diseased corn stalks in the fall to bury surviving spores.
6. Use resistant hybrids or varieties. Dent corn is generally more resistant than sweet or popcorn. In sweet corn, the larger, later-growing varieties usually are more resistant than the smaller, early varieties.

BACTERIAL SPOT, SPECK, AND CANKER OF TOMATOES

Bacterial spot, bacterial speck, and bacterial canker are widespread diseases of tomato that can cause localized epidemics during warm (spot and canker) or cool (speck), moist conditions. Bacterial spot can cause moderate to severe defoliation, blossom blight, and lesions on developing fruit. Bacterial speck also causes these symptoms but is usually not as severe as bacterial spot. Bacterial canker causes wilt, vascular discolouration, scorching of leaf margins, and lesions on fruit.

Symptoms

Foliar symptoms of bacterial spot and speck are identical. Small, water-soaked, greasy spots about 1/8 inch in diameter appear on infected leaflets. After a few days, these lesions are often surrounded by yellow halos and the centres dry out and frequently tear. Lesions may coalesce to form large, irregular dead spots. In mature plants, leaflet infection is most concentrated on fully-expanded and older leaves and some defoliation may occur. Spots may also appear on seedling stems and fruit pedicels. In some cases, blossom blight may occur, causing flower abortion. This is more severe with bacterial spot and may result in a split fruit set which is especially troublesome with determinate cultivars intended for mechanical harvest.

Bacterial spot and speck can usually be differentiated by symptoms on immature fruits. Bacterial spot lesions are small water-soaked spots that become slightly raised and enlarged until they are about 1/4 inch in diameter. The centres of these spots later become irregular, light brown, slightly sunken with a rough, scabby surface. In the early stages of infection, a white halo may surround each lesion at which time it resembles the fruit spot of bacterial canker. Small lesions which have not yet become scabby are often confused with lesions of bacterial speck. Bacterial speck appears on immature fruit as a black, slightly sunken stippling, eventually causing lesions less than 1/16 inch in diameter. Fruit lesions are not initiated on mature fruit in either disease.

Primary or systemic symptoms of bacterial canker (from infections originating in seeds or young seedlings) include stunting, wilting, vascular discolouration, development of open stem cankers, and fruit lesions. When affected stems are split open lengthwise, a thin, reddish-brown discolouration of the vascular tissue is observed, especially at the base of the plant. On young seedlings in the greenhouse, lesions may appear as raised pustules on leaves and stems. These plants rarely survive the season in the field. Secondary symptoms in the field include leaf "firing" (necrotic marginal leaf tissue adjacent to a thin band of chlorotic tissue) and fruit lesions. Spots on fruit are relatively small (1/32 to 1/16 inch) surrounded by a white halo ("bird's-eye" spots). Canker bacteria may also invade internal fruit tissues, causing a yellow to brown breakdown.

Causal Organisms

Bacterial spot is caused by the bacterium, Xanthomonas campestris pv. vesicatoria, which can be carried as a contaminant on the surface of infested seed and has been found to overwinter in soil associated with plant debris. Bacterial speck is caused by another bacterium, Pseudomonas syringae pv. tomato. This bacterium may also be seedborne and can overwinter on plant debris in soil and on the roots of many perennial plants. Bacterial canker is caused by Clavibacter michiganensis subsp. michiganensis, which, unlike the spot and speck pathogens, has the ability to infect tomato plants systemically. It is seedborne and can survive on infested plant debris in soil.

All three organisms may exist at low populations on leaf surfaces of symptomless plants. At the onset of favourable conditions, these low populations can increase rapidly and bacteria can then enter plants through stomata or small wounds and begin infection. Bacteria can spread rapidly with spattering rain and widespread epidemics may develop. Penetration of tomato fruit occurs through wounds created by windblown sand, breaking of hairs, or by insect punctures. Optimal conditions for bacterial spot and canker are high moisture, high relative humidity and warm temperatures (75 to 90 degrees F). Bacterial speck is more likely to occur under cool (64 to 75 degrees F), moist conditions.

1. Rotate tomatoes with non-solanaceous crops with at least 2 to 3 years between tomato crops. Avoid rotation with peppers, which are also susceptible to bacterial spot.
2. Plant only seed from disease-free plants or seed treated to reduce any bacterial populations. Treatments include:
   a. Fermentation of tomato pulp and seeds at room temperature for 4–5 days;
   b. Soaking seeds in 0.6–0.8 per cent acetic acid for 24 hr at 70 degrees F;

c. Soaking seeds 5ñ10 hr in 5 per cent hydrochloric acid;

d. Hot water treatment of seeds (122 degrees F for 25 minutes); or

e. Sodium hypochlorite (bleach) treatment [20–40 minute soak of seeds in 1 per cent sodium hypochlorite (20 per cent bleach)]. Some decrease in germination may be expected from these treatments.

3. Use only transplants free of disease symptoms.

4. Carry out proper sanitation of transplant production greenhouses. Remove all weeds and plant debris, clean all tools with disinfectant solution, and wash hands thoroughly before and after handling plants. Water plants early in the day to reduce the amount of time foliage is wet. Do not handle plants when they are wet. After each crop, clean greenhouse walls, benches, etc., with hot soapy water, followed by thorough rinsing and treatment with a disinfectant. If possible, close up greenhouse after transplant production is completed to allow natural heating during the summer. Use only new plug trays and pathogen-free planting mixes. Avoid growing peppers and tomatoes in the same greenhouse unless pepper seed has also been treated as in step 2.

5. In the field, control irrigation to minimize the time foliage is wet and avoid working among wet plants to minimize spread of disease.

6. Applications of mancozeb plus copper soon after transplanting may help retard development and spread of bacterial spot and speck. This practice is not particularly effective for management of bacterial canker. Many tomato processors will not accept tomatoes treated with mancozeb or other EBDC fungicides. Check with your processor before applying one of these fungicides.

BACTERIAL WILT OF CUCURBITS

After vine crops begin to run, gardeners and farmers often notice individual leaves with severe wilt symptoms on sunny days. Within a week or two the condition spreads to entire vines which do not recover from the wilt. This disease, called bacterial wilt, is especially common with cantaloupes and cucumbers. Squash and pumpkins may not wilt as rapidly, but may be dwarfed with extensive blossoming and branching. Watermelons are rarely affected.

Symptoms

Wilting of individual leaves or vines of the plant is the characteristic symptom. One or a few leaves wilt and become dull green. The disease spreads from the leaves downward into the petioles and then the stem until the entire plant wilts and dies.

There are other factors, such as vine borers and soil-borne fungal pathogens, that may cause cucurbits to wilt. Sometimes, if an affected stem is cut off near the ground, the sap may be milky in appearance or sticky and, if touched with the finger, the sap will string up to half an inch. This is a helpful test in diagnosis of bacterial wilt, but cannot be depended upon for positive identification.

Causal Organism

This disease is caused by a bacterium, Erwinia tracheiphila, that overwinters in the bodies of the striped and 12-spotted cucumber beetles. In the spring, the beetles emerge from the ground and feed on young plants, introducing bacteria into the leaves or stems. The bacteria reproduce in the water-conducting vessels, producing gums that interfere with water transport.

The beetles and bacteria are so intimately related that controlling the beetles will control infection by the bacteria. Once infection has occurred, however, no control is possible and wilting plants should be removed, if practical.

Management

The only practical management measure is to use an insecticide when seedlings first emerge to control the black and yellow cucumber beetles. Early infections are most severe, but total control depends on applications continuing at frequent intervals during the growing season. In some cases, if insect pressure is heavy, it may be necessary to apply an insecticide when plants are just cracking the soil, but have not yet emerged. Management of this disease is completely linked with preventing feeding of cucumber beetles on susceptible hosts.

FUSARIUM AND VERTICILLIUM WILTS OF TOMATO, POTATO, PEPPER, AND EGGPLANT

Solanaceous crop plants (tomato, potato, pepper, and egg-plant) may be infected at any age by the fungi that cause Fusarium wilt and Verticillium wilt. The wilt organisms usually enter the plant through young roots and then grow into and up the water conducting vessels of the roots and stem. As the vessels are plugged and collapse, the water supply to the leaves is blocked. With a limited water supply, leaves begin to wilt on sunny days and recover at night. Wilting may first appear in the top of the plant or in the lower leaves. The process may continue until the entire plant is wilted, stunted, or dead.

Tomato and potato plants may recover somewhat but are usually weak, unthrifty, and produce fruit of low quality. Peppers typically collapse rapidly and die. Fusarium and Verticillium wilts are rarely significant in field grown tomatoes due to the widespread incorporation in tomato cultivars of genes for resistance to the pathogen. However, the resurgent interest in planting "heirloom" tomato varieties which do not carry resistance genes has resulted in increased incidence of Fusarium and Verticillium wilts. Additionally, new races of both pathogens have been identified that are capable of overcoming the resistance in many popular tomato varieties. Verticillium race 2 is now common in tomatoes, but its importance in reducing yield is not known at this time. There is very little genetic resistance available to either disease in pepper or eggplant.

Symptoms

Fusarium Wilt

Fusarium wilt symptoms begin in tomato and potato as slight vein clearing on outer leaflets and drooping of leaf petioles. Later the lower leaves wilt, turn yellow and die and the entire plant may be killed, often before the plant reaches maturity. In many cases a single shoot wilts before the rest of the plant shows symptoms or one side of the plant is affected first. If the main stem is cut, dark, chocolate-brown streaks may be seen running lengthwise through the stem. This discolouration often extends upward for some distance and is especially evident at the point where the petiole joins the stem. Potato tubers may show browning of the vascular ring as well as browning at the stem end and decay where stolons are attached. In pepper, lower leaves do not begin to wilt until roots and the base of the stem have already started to decay. Wilting of the entire plant soon follows. Dark brown, sunken, and eventually girdling cankers may be seen at the base of the pepper plant. In eggplant, wilting progresses from lower to upper leaves, followed by collapse of the plant.

Verticillium Wilt

Verticillium wilt symptoms on tomato, potato, and eggplant are similar to those of Fusarium wilt. Often no symptoms are seen until the plant is bearing heavily or a dry period occurs. The bottom leaves become pale, then tips and edges die and leaves finally die and drop off. V-shaped lesions at leaf tips are typical of Verticillium wilt of tomato. Infected plants usually survive the season but are somewhat stunted and both yields and fruits may be small depending on severity of attack.

A light tan discolouration in the stem similar to that caused by Fusarium wilt may be found but is usually confined to lower plant parts. The discolouration is typically lighter in colour than with Fusarium wilt. Symptoms on one side of the plant only are sometimes seen. In potatoes the pathogen may be part of a complex that includes, among others, the root lesion nematode and the bacterial soft rot organism, resulting in premature plant death ("potato early dying disease"). Tubers from Verticillium-infected plants may show light brown vascular discolouration, usually restricted to the stem end. In pepper, the lower leaves wilt, then leaf tips and margins dry and turn brown. Brown streaks in the vascular tissue can be observed well up into the plant, which rapidly collapses and dies.

Causal Organisms

Fusarium wilt in solanaceous crops is caused by several different types of the fungus Fusarium oxysporum. These are: F. oxysporum f. sp. lycopersici (tomato), F. oxysporum f. sp. melongenae (eggplant) and F. oxysporum var. vasinfectum (pepper). Fusarium wilt in potato is caused by a complex of up to four different Fusarium spp. All of the Fusarium wilt pathogens are generally specific to their hosts and are soilborne. They are warm weather organisms, and therefore Fusarium wilts are most serious later in the growing season.

Verticillium wilt is caused by the fungi Verticillium albo-atrum and V. dahliae. These fungi attack a wide range of plant species, including cultivated crops and weeds. They are soilborne in field and greenhouse soils where they can persist for many years. V. albo-atrum is a cool weather organism that grows best when soil temperatures are between 65 and 75 degrees F. V. dahliae is more active between 75 and 83 degrees F. Although disease is retarded by the higher temperatures that favour Fusarium wilt, visible symptoms may appear to be more severe when high temperatures exist, due to restricted water movement in the plant brought about by damage done to the water conducting vessels earlier in the growing season.

1. Because Fusarium and Verticillium fungi are widespread and persist several years in soil, a long crop rotation (4 to 6 years) is necessary to reduce populations of these fungi. Avoid using any

solanaceous crop (potato, tomato, pepper, eggplant) in the rotation, and if Verticillium wilt is a problem, also avoid the use of strawberries and raspberries, which are highly susceptible. Rotate with cereals and grasses wherever possible.

2. Keep rotational crops weed-free (there are many weeds hosts of Verticillium).
3. Whenever practical, remove and destroy infested plant material after harvest.
4. Maintain a high level of plant vigour with appropriate fertilization and irrigation, but do not over-irrigate, especially early in the season.
5. Plant disease resistant tomato varieties, labeled V (for Verticillium) and F (for Fusarium). These disease resistance designations are usually shown in seed catalogues. Fusarium- or Verticillium-resistant varieties of eggplant, potato, and pepper are generally not available.
6. If soils are severely infested, production of solanaceous crops may not be possible unless soil fumigation is an option.

BACTERIAL SPOT OF PEPPER

Bacterial spot of pepper is one of the most destructive diseases of pepper in climates where high temperature and frequent rainfall occur during the growing season. The disease causes spots on leaves and fruit; leaf defoliation; and a reduction in plant growth, fruit yield, and quality. Bacterial spot is also a serious problem on tomato, although not all strains of the pathogen can cause disease on both hosts.

Symptoms

Characteristic bacterial spot symptoms can appear on the leaves, fruits, stem, and petioles. On leaves, symptoms begin as small, yellow-green circular lesions surrounded by a yellowish halo. These spots appear water-soaked under wet conditions. As the lesions mature, a general yellowing extending from the area around the lesions develops on diseased leaves and the centre of the spots become brown to black and sunken.

Tissue in the centre of the lesion often dries and breaks away, giving a "shot-hole" appearance to the leaf. When spots are numerous, they may join together and form irregular discoloured streaks along the veins and leaf margins. Edges and tips of leaves may die, then dry and break away, causing leaves to appear ragged. Severely spotted leaves turn yellow or brown and fall from the plant; young leaves can be distorted. Fruit spots begin as green, circular, slightly raised lesions which eventually become brown or dark, raised, and about 1/8 inch in diameter. Centres of the spots become necrotic, corky, and scab-like.

On stems and petioles, lesions are elongated and blackened, and can kill leaflets. Cotyledons are particularly susceptible to bacterial spot; lesions are initially small, sunken, and silvery. They later become darker in colour.

Causal Organism

Bacterial spot of pepper is caused by a bacterium, Xanthomonas campestris pv. vesicatoria, which can overwinter in crop residue in or on soil, on or in seeds, and on wild host plants. Pepper seeds infested with the pathogen are a major source of inoculum for bacterial spot as well as the major means of long distance spread of the pathogen. The pathogen can survive on dried seeds for up to 10 years. It is not able to survive free in soil for a long time, but can survive up to 6 months in infected crop debris in soil. The bacterium penetrates leaves through stomata and/or wounds, and fruits through wounds created by wind-driven sand, insect punctures, or mechanical injury. Dissemination of the bacterium occurs between and within fields by water-splashing, aerosols, or during cultivating, hoeing, thinning of direct seeded plants, transplanting, or harvesting.

In the United States, numerous physiological races of the bacterium are known, most of which have been found.

1. Use pathogen-free seeds and transplants.
2. Use sodium hypochlorite-treated seed to reduce bacterial populations.
3. Practice crop rotation with non-host plants such as corn and soybean so that peppers are grown only every 3 to 4 years. However, do not use soybeans in the rotation if white mold (Sclerotinia sclerotiorum) has been a problem.
4. Deep plow to bury infected crop debris.
5. Avoid working in the field when foliage is wet.
6. Eliminate wild host plants such as nightshade and ground cherry in and around field.
7. Application of copper-containing pesticides may be helpful for preventing development and spread of bacterial spot.
8. None of the currently available pepper varieties are resistant to all known races of the bacterial spot pathogen. However, use of varieties resistant to one or more races of the pathogen may provide some control, depending on the races present.

DOWNY MILDEW OF CRUCIFERS

Downy mildew affects all cultivated plants and weeds in the crucifer family. It can be a serious problem in commercial production of cabbage, broccoli, cauliflower, radish, turnip, mustard, collard, and cruciferous greens. Under favourable conditions, it may cause serious losses in the field or may develop after harvest and cause deterioration of product quality during packing and shipping.

Symptoms

Plants can be infected at any stage of development. In seed beds, cotyledons and primary leaves are invaded resulting in fungal growth visible on the underside of the leaf. Later a slight yellowing develops opposite the fungal growth on the upper side of the leaf. The young leaf or cotyledon, when yellow, may drop off. Older leaves usually persist and infected areas gradually enlarge, turn bright yellow, then become tan and papery. Rarely the affected leaf may develop hundreds of minute darkened specks. Under cool, moist conditions, a white mildew growth can be seen on the underside of infected leaf lesions.

Symptoms may appear on other plant parts as well. The fleshy roots of turnips and radishes may develop an internal, irregularly shaped discolouration extending from the crown downward. The flesh may be brown or black or show a form of net necrosis. In advanced stages, the skin becomes roughened by minute cracks and the root may split open. In radish these symptoms may be confused with those caused by Rhizoctonia.

Causal Organism

The fungus causing downy mildew in crucifers, Peronospora parasitica, overwinters in roots or in decaying portions of diseased plants. Thick-walled resting spores may form in stems, cotyledons, and other fleshy parts of infected host plants. On growing plants, the fungus produces large numbers of spores that are blown about by wind and splashed by rain. Moisture and temperature are important in the spread and reproduction of this fungus. High relative humidity during cool or warm, but not hot, periods promotes its growth and sporulation. Presence of a water film on the foliage from fog, drizzling rain, or dew allows spores to germinate, infect, and produce more spores on a susceptible host in as few as 4 days.

1. Use a crop rotation plan that excludes production of any type of cruciferous crop for at least 2 out of every 3 years.
2. Practice sanitary measures such as the use of clean

seed beds away from other crucifer production and the destruction of cruciferous weeds.

3. Use a planting site and plant spacing pattern that expose plants to full sun throughout the day.
4. If severe disease pressure is expected, apply a registered fungicide weekly beginning soon after emergence.
5. Disease resistant cultivars are not available for most cruciferous crops. However, some hybrid cultivars of broccoli are resistant or tolerant to downy mildew.

BLACK ROT OF CRUCIFERS

Black rot is the most serious disease of crucifer crops world wide when environmental conditions (relatively high temperature and humidity) are favourable. The disease affects primarily aboveground parts of plants at any stage of growth and causes high yield and quality losses, especially in tropical and subtropical regions during the rainy season. All vegetables in the crucifer family, including broccoli, Brussels sprouts, cabbage, cauliflower, Chinese cabbage, kale, mustard, radish, rutabaga, and turnip, are susceptible to black rot. Many cruciferous weeds such as Shepherd's Purse, wild mustard, and yellow rocket may also be hosts of this pathogen.

The characteristic black rot symptom on most cultivated crucifer plants is the appearance of yellow, V-shaped lesions along the margins of leaves. The point of the V-shaped lesion is directed toward a vein. When lesions enlarge, wilted tissue expands toward the base of leaves. Eventually the diseased areas become necrotic and the veins turn black or brown.

The infection may move down the vascular tissue of petioles and then spread up and down the stems. When stems and petioles of an infected plant are cut crosswise or lengthwise, the black-brown vascular tissue with yellowish bacterial slime is observed. These symptoms may be confused with Fusarium yellows, except that Fusarium causes brown vein discolouration without bacterial slime. Moreover, symptoms of black rot may vary according to age of host, host genus, species, and cultivar and even environmental conditions. Symptoms on cauliflower may appear as numerous black or brown specks, scratched leaf margins, black veins, and discoloured curds. Many cruciferous weed species do not exhibit any of these characteristic symptoms even when infected.

Symptoms

Causal Organism

Black rot of crucifer is caused by a bacterium, Xanthomonas campestris pv. campestris (Xcc). The bacteria can overwinter in plant debris, in and on seeds from diseased plants, and in and on weeds. The pathogen may survive in diseased crop residue buried in soil for up to 2 years, but not more than 60 days free in soil. The major source of these bacteria is infected seeds, which enable long-distance spread of the disease. The pathogen is spread within and between fields by splashing water, wind, insects, machinery, and irrigation or drainage waters. The bacteria infect the cotyledons and young leaves through natural plant openings (stomata, hydathodes) or wounds and then migrate between cells until they reach the xylem tissue where they spread throughout the plant. Free moisture is required for infection by the pathogen. After infection, symptoms may appear on plants within 7 to 14 days under optimum conditions (25 to 30 degrees C).

Effective management of black rot of crucifers depends on the application of the following practices in combination:

1. Use black rot-tested, disease-free seed grown in an arid production area.
    a. If source of the seeds is unknown, or infested seedlots must be used, treat seed with hot water to eradicate pathogenic bacteria. Cabbage, broccoli, and Brussels sprouts can be treated at 50 degrees C for 25 minutes, while seeds of cauliflower, kale, turnip, and rutabaga are treated for 15 minutes. However, this treatment may reduce the viability of seed. Therefore, some

other chemical seed treatments, including, sodium hypochlorite, hydrogen peroxide, and hot acidified cupric acetate or zinc sulfate can be applied to eliminate the bacteria from crucifer plant seeds.

2. Use certified disease-free transplants.
3. Practice crop rotation where crucifers are grown only every 3 to 4 years to eliminate the inoculum sources from diseased crop debris in the soil.
4. Good sanitation practices should be performed to prevent disease spread.
   a. Eliminate all volunteer crucifer plants from previous crops and alternative wild host plants within and around the field.
   b. Do not apply manure that may contain crucifer residues.
   c. Do not use sprinkler irrigation.
   d. Avoid working in the field when plants are wet.
   e. Do not allow machinery and equipment movement from infested areas to non-infested fields.
   f. Deep plow to bury all crucifer residues after harvest.
5. Application of fixed copper pesticides in the field may help to reduce spread of the disease.
6. A few black rot-resistant cultivars of cabbage and other crucifers are commercially available. These resistant cultivars should be used in crucifer growing regions where black rot is a common problem.

GUMMY STEM BLIGHT AND BLACK ROT OF CUCURBITS

Gummy stem blight is an important disease of squash, pumpkins, cucumbers, watermelons and other field-grown cucurbit crops. It can occur at any growth stage, from seedlings to mature plants. This disease on fruits, in the field or in storage, is called black rot. The disease also can cause extensive damage to all above-ground parts of greenhouse-grown cucumbers.

Symptoms

Gummy stem blight occurs on all plant parts except roots. Leaf symptoms appear as dark yellow or reddish-brown lesions in various shapes. Lesions begin at leaf margins and extend rapidly back into the leaf blade, causing curling, shriveling, and death of the entire leaf. Pimple-like structures (pycnidia) may be found in leaf lesions by close inspection with a hand lens.

Fruit symptoms vary among crops. Winter squash (Hubbard, butternut, etc.) are likely to show symptoms primarily on fruit or older leaves. Black rot symptoms appear as a brown to black rot of the rind, flesh, and seed cavity accompanied by heavy white and black fungus growth. Lesions may develop anywhere on the fruit, first as water-soaked areas dotted with pycnidia that ooze yellowish masses of spores. On Hubbard-type squash, the brownish-black rot extends down into the flesh and seed cavity. Seeds often become dotted with small black pycnidia. On butternut squash, lesions are brown with irregular ring patterns and are superficial over the skin surface, not penetrating into the flesh or seed cavity.

Infection usually occurs in the field, causing water-soaked, cracked, brownish cankers on the vines. A reddish gum may develop in these cracks, although this alone is not a diagnostic sign. (Fusarium and scab may also produce a reddish gum.) Fruit may decay at the site of attachment as a result of the fungus invading the stem. Butternut squash and gourds may develop black rot before harvest, but Hubbard squash are resistant during growth and do not show symptoms until the storage period. Fruit rot on greenhouse cucumbers usually begins at the blossom end of immature fruit. Lesions are firm and become dark brown to black when cut open. Occasionally, lesions develop on one side of a fruit, causing it to hook as it grows.

Causal Organism

Gummy stem blight is caused by the fungus Didymella bryoniae. The pathogen can be seed-borne or can survive on organic debris from previously infected cucurbits or on wild or volunteer cucurbits. The gummy stem blight fungus produces two types of spores. Windblown ascospores are likely to start the disease in a field. Later, pycnidiospores are released in a gummy substance that makes them adaptable for short distance spread by splashing water. Spore production and infection are influenced by moisture and temperature. A moisture film from dew, rain, or overhead irrigation is necessary for spore germination.

Optimum temperature for infection is 61 to 75 degrees F. Low night temperatures, particularly in greenhouses, may cause water droplets to exude from leaf points and condensation to form on leaves, favouring infection by the fungus at those points. Infection of fruits commonly occurs through harvest wounds or through dying flowers.

1. Use only disease-free seed produced in arid western locations.
2. Plow crop refuse deeply to reduce survival of the fungus as soon as crop is harvested.
3. Practice crop rotation with non-cucurbit crops so that cucurbits are grown only every 3 to 4 years.
4. Apply foliar-protectant fungicides on a routine basis.
5. Avoid wounding fruit during harvest and store fruit at 45 to 50 degrees F to prevent postharvest fruit rots.
6. Resistant cultivars are not currently available.

MUMMY BERRY OF BLUEBERRY

Mummy berry is one of the most serious diseases of blueberry. Once the disease becomes established in a planting, it can destroy most of the crop. Losses result from:

1. Rotted berries; and
2. Killing or blighting of blossoms, blossom and leaf clusters and young shoots.

Symptoms

Blighting of new shoot tips and blossoms can be easily mistaken for frost damage. By blossom time, the infected young leaf and shoot growth will wilt, turn brown and die. About a week or so after infection of the early new growth occurs, dead areas develop on the petioles and along the midrib and veins of the leaves or at the base of flowers. Berries that develop from infected flowers may attain nearly full size before turning tan or gray and shriveling into hard mummies, which drop to the ground at or before harvest.

Causal Organism

Mummy berry is caused by the fungus, *Monilinia vaccinii-corymbosi*. The fungus overwinters in the shriveled mummies on the ground. In early spring, cup- or globe-shaped structures of the fungus called apothecia are produced on mummified berries during cool rainy periods. Spores produced inside apothecia are released into the air and carried by the wind to young developing leaf shoots and flowers where they cause primary infections. If moisture is not present, the fungus will not produce spores. However, the fungus may survive in mummies for one year or more. Another type of fungus spores (conidia) is produced on dead tissue that results from primary infections. These conidia are spread during bloom by wind and insects, and result in secondary infection of flowers. Fruit that develops from infected flowers turns into mummies and falls to the ground.

Control

Removing mummified berries from the planting will greatly aid in controlling the disease. Removing these berries is not practical on a commercial scale, but may be of value in backyard plantings where there are just a few plants. After removing mummies, burn or bury them.

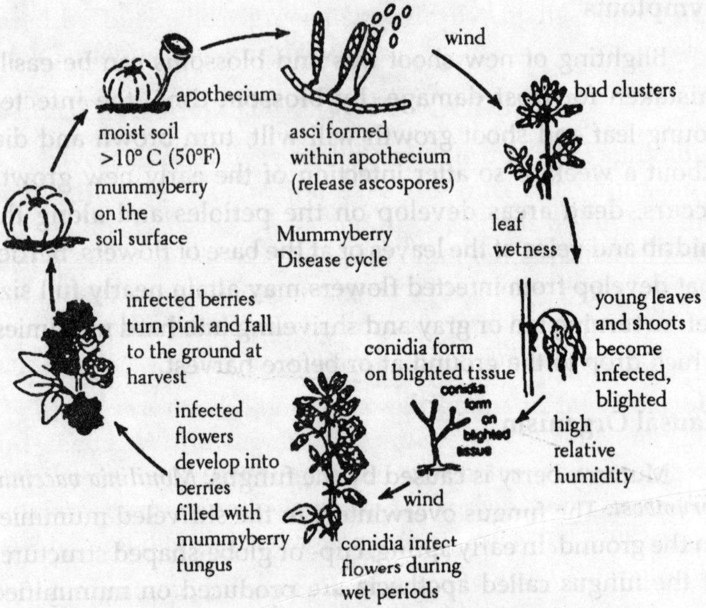

Fig. Disease cycle of mummy berry

Cultivation in early spring to disturb or cover the mummies has been reported to be effective. Mummies that are disturbed or covered with soil at this time remain dormant or do not produce spores. Cultivation between rows and raking under plants to disturb or cover mummies should be done as early as possible in the spring and repeated after each hard rain until after bloom. If just a few mummies are missed, they can produce enough spores to infect the planting.

Where mummy berry is a problem, a good fungicide spray programme is essential.

LEATHER ROT OF STRAWBERRY

Leather rot of strawberry has been reported in many regions of the United States. In many areas, it is considered a minor disease of little economic importance. However, excessive rainfall during May, June, and July can lead to severe losses in fruit yield and quality resulting from leather rot. Commercial growers have lost up to 50 per cent of their crop to leather rot. The leather rot fungus primarily attacks the fruit, but many also infect blossoms.

Symptoms

Leather rot can infect berries at any stage of development. Where the disease has been a problem, infection of green fruit is common. On green berries, diseased areas may be dark brown or natural green outlined by a brown margin. As the rot spreads, the entire berry becomes brown, maintains a rough texture, and appears leathery. The disease is more difficult to detect on ripe fruit. On fully mature berries, infection may result in little colour change or discolouration ranging from brown to dark purple. Infected ripe fruit are usually softer to the touch than healthy fruit. When diseased berries are cut crosswise, a marked darkening of the water-conducting system to each seed can be observed. In later stages of decay, mature fruits also become tough and leathery. Occasionally, a white moldy growth can be observed on the surface of infected fruit. In time, infected fruit dry up to form stiff, shriveled mummies.

Berries that are affected by leather rot have a distinctive, unpleasant odor and taste. Even the healthy tissue on a slightly rotted berry is bitter. This presents a special problem to growers in pick-your-own operations. An infected mature berry with little colour change may appear normal and be picked and processed with healthy berries. Consumers have complained of bitter tasting jam and jelly made with berries from fields where leather rot was a problem. Leather rot is observed most commonly in poorly drained areas where there is or has been free standing water, or on berries in direct contact with the soil.

Causal Organism

Leather rot is caused by the fungus, *Phytophthora cactorum*. The fungus survives the winter as thick-walled, resting spores, called oospores, that form within infected fruit as they mummify. These oospores can remain viable in soil for long periods of time. In the spring, oospores germinate in the presence of free water to produce a structure called a sporangium.

A second type of spore, called a zoospore, is produced inside the sporangium. Up to 50 zoospores may be produced inside one sporangium. The zoospores have tiny flagella and can swim in a film of water. In the presence of free water on the fruit surface, the zoospores germinate forming a germ tube and infect the fruit. In later stages of infection, sporangia are produced on the surface of infected fruit under moist conditions. The fungus is spread by splashing or windblown water from rain or overhead irrigation. Sporangia and/or zoospores are carried in water from the surface of infected fruit to healthy fruit where new infections occur. Under the proper environmental conditions, the disease can spread very quickly. A wetness period (free water on fruit surface) of one hour is sufficient for infection. The optimum temperatures for infection are between 62 and 77 degrees F (17-25 degrees C). As the length of the wetness period increases, the temperature range at which infection can occur becomes much broader. As infected fruit dry up and mummify, they fall to the ground and lie at or slightly below the soil surface. Oospores formed within the mummified fruit enable the fungus to survive the winter and cause new infections the following year, thus completing the disease cycle.

Control

- Select a planting site with good soil drainage and air circulation. Good soil drainage is critical. Sites that drain poorly or are subject to periodic flooding are ideal for leather rot development.
- Plants should be exposed to direct sunlight (avoid shade). Plant rows with the direction of the prevailing wind to promote faster drying of foliage and fruit.
- Mulch strawberry plants with straw or other material that reduces fruit contact with soil. Research has shown that a good layer of straw mulch is very beneficial in controlling leather rot.
- Proper spacing of plants and timing of fertilizer applications are important. Excessive applications of

nitrogen fertilizer can produce excessive amounts of dense foliage. Shading of berries by thick foliage prevents rapid drying of fruit during wet periods and creates ideal conditions for disease development.

- Pick fruit frequently and early in the day (as soon as plants are dry). Cull out all diseased berries, but do not leave them in the field.

CANE BLIGHT OF RASPBERRIES

Cane blight is one of the more damaging diseases of raspberries. The disease is most common on black raspberries but also occurs on red and purple varieties. The disease occasionally occurs on blackberries and dewberries. Cane blight can result in wilt and death of lateral shoots, a general weakening of the cane, and reduced yield. It is usually most severe during wet growing seasons.

Symptoms

On first year canes (primocanes) dark brown-to-purplish cankers form on new canes near the end of the season where pruning, insect, and other wounds are present. The cankers enlarge and extend down the cane or encircle it, causing lateral shoots above the diseased area to wilt and eventually die. Black specks, which are reproductive bodies of the cane blight fungus develop in the brown cankered bark. In wet weather, large numbers of microscopic spores ooze out of the pycnidia. This ooze gives the bark a dark-gray, smudgy appearance. During winter, infected canes commonly become cracked, brittle, and snap off easily. On infected second-year canes (floricanes), the side branches may suddenly wilt and die, usually between blossoming and fruit ripening. Upon close examination, the presence of dark brown or purplish cankers can be observed on the main cane or branches below the wilted area.

Causal Organism

Cane blight is caused by the fungus, *Leptospaeria coniothyrium*. The pathogen survives over winter on infected or dead canes. The following spring, spores are released and carried by splashing rain and wind to nearby primocanes. Under moist conditions, the spores germinate and penetrate pruning wounds, insect punctures, fruit stem breaks and other wounds. After entry the fungus rapidly invades and kills bark and other cane tissues. Fungal fruiting bodies are formed in older cankers and complete the disease cycle. Dead canes continue to produce conidia and remain a source of infection for several years.

Control

- All steps possible should be taken to improve air circulation within a planting, to allow faster drying of foliage and canes. Reducing the number and duration of wet periods should reduce the potential for infection. Excessive applications of fertilizer (especially nitrogen) should be avoided, since it promotes excessive growth of very susceptible succulent plant tissue. Plants should be maintained in narrow rows and thinned to improve air circulation and allow better light penetration. Weeds are very effective in reducing air movement; therefore, good weed control within and between rows is important for improving air circulation within the planting. Raspberries should be planted in sunny, open areas where water and air drainage are good. This allows plants to dry quicker after wet periods, and reduces the chance of infection.
- Wild brambles, especially wild raspberries, growing in the area should be removed. They can provide a continuous source of spores to spread this and other diseases and pests to cultivated raspberries and blackberries.
- Healthy, rapidly growing plants that have been properly fertilized and watered, are more resistant to cane blight.

- After harvest, remove and destroy all old fruited floricanes and any new primocanes canes that are infected. Old canes should be removed before growth starts in the spring.
- Keep plantings free of insects, since they may cause wounds that serve as entry points for the fungus. Avoid any other pests or cultural practices that result in wounding of the canes.
- If cane blight is a serious problem, the use of fungicides should be considered.

EUTYPA DIEBACK OF GRAPE

For more than 60 years the eastern grape industry recognized a disease called "dead-arm" thought to be caused by the fungus *Phomopsis viticola*. In 1976, researchers demonstrated that the "dead-arm" disease was actually two different diseases that often occur simultaneously: Eutypa dieback and Phomopsis cane and leaf spot. Eutypa dieback is the new name for the trunk and arm phase of what was once known as "dead-arm."

Scientists now propose that the name "dead-arm" be dropped. It is important that Eutypa dieback and Phomopsis cane and leaf spot be considered as two distinctly different diseases because the control recommendation for each are quite different.

Symptoms

The earliest symptom is a canker that generally forms around pruning wounds in older wood of the main trunk. These cankers are usually difficult to see because they are covered with bark. One indication of a canker is a flattened area on the trunk. Removal of bark over the canker reveals a sharply defined region of darkened or discoloured wood. These cankers may be up to 3 feet long, but do not enter one- or two-year wood and seldom go below ground line. When the trunk is cut in cross-section, the canker appears as darkened or discoloured wood extending in a wedge shape to the centre of the trunk.

The most striking and obvious symptoms of Eutypa dieback are the leaf and shoot symptoms, which may not develop for up to 3 years after infection of the vine has occurred. These symptoms are most obvious in spring (May and June). Spring shoot growth is weak and stunted above the cankered area. Leaves are at first smaller than normal, cupped, misshapen and yellowed. Later in the season (mid July), these leaf and shoot symptoms may disappear from all but the basal leaves of affected shoots. The vines may appear to have recovered. However, the infected trunk and all growth above it will eventually die.

CAUSAL ORGANISM

Eutypa dieback is caused by the fungus, *Eutypa armeniacea.* The fungus survives in infected trunks for long periods of time, whether they remain as part of the in-place vine or as prunings in the vineyard. Eventually, the fungus produces reproductive structures (perithelia) on the surface of infected wood. Spores (ascospore) are produced in these structures and are discharged into the air. Ascospore discharge is initiated by the presence of free water (either rainfall or snow melt). Most spores appear to be released during winter or early spring, with relatively few being released during the summer months. Unfortunately, most spores are released at the same time pruning is being conducted. The ascospore can be carried considerable distances by air currents to recent wounds on the trunk. Pruning wounds are by far the most important points of infection. When the ascospore come in contact with newly cut wood, they germinate and a new infection is initiated.

Control

1. At present the primary control methods is removing infected trunks from the vineyard. The vine must be cut off below the cankered or discoloured wood. If the canker extends below the soil line, the stump can be left and a new trunk formed. It is important to remember that the best time to identify infected vines is spring (May and June) and when the leaf and shoot

symptoms are most obvious. If trunks cannot be removed in the spring, they should at least be marked so they can be removed after harvest, but before the next spring.

2. Sanitation is critical. All wood from infected plants must be removed from the vineyard and destroyed (either buried or burned). An old infected trunk lying on the ground may continue to produce spores for several years.
3. At present, no fungicide recommendations are available for control of this disease.

NECROTIC LEAF BLOTCH OF GOLDEN DELICIOUS APPLES

Necrotic leaf blotch is a common but minor disease that appears to be restricted to the Golden Delicious cultivar of apples and its sports. The disease is a physiological dis-order and has been observed for about 20 years in most apple-growing areas of the Eastern and Midwestern United States.

Necrotic leaf blotch usually appears in late June or early July and is most severe in the latter part of the growing season. Mature leaves from the base to the centre of upright, succulent, rapid-growing shoots are usually the only ones that show symptoms. Young, immature leaves on succulent shoots and mature or cluster leaves on fruiting wood are usually not affected.

The effect of the disease varies among individual Golden Delicious trees, both from orchard to orchard and from tree to tree within the same orchard.

Symptoms

The symptoms are a leaf blotching, followed by a yellowing of the affected leaves (chlorosis) and defoliation. Defoliation can be severe. Irregular brown blotches, usually 0.5 to 1.5 centimeters in diameter, suddenly appear on leaves. The blotches are usually restricted by the larger leaf veins. Many affected leaves turn yellow and drop all at once a few days after the necrotic blotches first appear. The most conspicuous symptom of necrotic leaf blotch is the large number of yellow leaves, on the tree and later on the ground, that suddenly appear during July and August.

A distinctive characteristic of the disease is its appearance in two to four, or more, waves of these symptoms during certain periods that are scattered throughout the latter half of the growing season. Between these periods or waves, little or no additional disease develops. During the course of the season, 10 to 50 per cent defoliation may occur on severely affected trees. All sports of Golden Delicious on any root stock appear to be equally susceptible.

Cause

Necrotic leaf blotch of Golden Delicious apples appears not to be caused by a fungus, bacterium, air pollutant, or a nutrient deficiency. The causal agent of necrotic leaf blotch is unknown; therefore, the disease is often referred to as a physiological disorder. The disease commonly appears during hot, hazy weather following a humid, rainy period. Other environmental factors or an air pollutant may be involved. The levels of ozone or sulfur dioxide in the air and foliar nutrients in the leaves have little or no effect on disease development.

Control

1. It has been observed that certain fungicides result in less necrotic leaf blotch. Since a fungal pathogen does not appear to cause necrotic leaf blotch, it is difficult to assess the role of fungicides in reducing necrotic leaf blotch. The effect may be indirect, by altering leaf physiology. Fungicides containing zinc (Zn) ions appear to be the most effective. Golden Delicious trees sprayed, during the summer cover spray period, with fungicides containing mancozeb (Dithane M-45, Manzate D, Penncozeb) or Ziram tend to have less necrotic leaf blotch than unsprayed trees or trees sprayed with other fungicides.

2. Trees that have an annual moderate crop of fruit have less necrotic leaf blotch than trees that have a biennial bearing habit or a light fruit crop.
3. Golden Delicious trees pruned to a central leader commonly have less necrotic leaf blotch than trees pruned to an open centre.

WHITE PINE BLISTER RUST ON CURRANTS AND GOOSEBERRIES

White pine blister rust is not a serious disease of currants and gooseberries; however, it is a very serious disease of white pines (*Pinus strobus*). Currants and gooseberries serve as an alternate host for the rust fungus that causes white pine blister rust. Therefore, planting currants and gooseberries in areas where white pines are present can lead to serious losses of white pines. North American white pine species, including bristlecone, limber, sugar, eastern white, southwestern white, western white, and whitebark, are highly susceptible. White pine blister rust causes significant damage in pine forests by forming cankers on the branches of white pines. These cankers ultimately kill the trees. Black currant is the most susceptible of the *Ribes* species.

To protect white pine forests, several states have enacted laws concerning planting of black currants.

The current law to suppress and control White Pine Blister Rust Disease is as follows:

1. The European black currant, *Ribes nigrum* L. or any variety of this species is hereby declared to be a public nuisance, and it shall be unlawful for any person to possess, transport, plant, propagate, sell, or offer for sale, plants, roots, scions, seeds, or cuttings of these plants in this state.
2. Recognized varieties, e.g., "Consort" produced by the hybridization of *Ribes nigrum* L. or a variety thereof with a resistant or immune species, known to be immune or highly resistant to the White Pine Blister Rust fungus.

Symptoms

On *Ribes* in the spring, tiny yellowish spots become visible on the upper surface of the leaves, while on the underside, orange-yellow blister-like fruiting bodies appear. By late summer, yellow to brown threadlike growths develop on or near these infection spots on the leaf. Bushes also will have premature defoliation.

On white pine, the symptoms include dead branches, chlorotic foliage, branch girdling by lesions that exude resin or sticky yellowish fluid (spermagonia), cankers that are diamond-shaped to elliptic with a dead centre surrounded by a band of yellowish-green infected bark, light yellow-orange aecia, and death of the tree.

Causal Organism and Disease Development

White pine blister rust is caused by a fungus, *Cronartium ribicola*. The organism was introduced from Europe in the early 20th Century. It has spread throughout the entire range of white pines in North America. The life cycle takes three to six years to complete. The initial infection of black currant bushes occurs in the spring, when aeciospores from diseased white pine land on the leaves of the bush. These spores can travel on the wind several miles. Moisture is needed for the germination of the aeciospores.

After one to three weeks incubation, the plants begin to show the first symptoms of disease. Yellowish spots appear on the top side of the leaves and fruiting bodies (uredia) appear on the underside of the leaves. The fruiting bodies (uredia) produce uredospores which reinfect *Ribes* leaves. This cycle can be repeated many times during a single growing season.

In late summer or early fall, when day length and temperature decrease, telia (another type of fruiting body) replace the uredia on the underside of the leaves. The telia produce teliospores, which germinate in place at night during wet weather and produce basidiospores. After telial formation, at least 60 hours of wet weather with temperatures not exceeding 20C are required for basidiospore formation, dispersal, and infection of white pine. These basidiospores are carried by the air currents to white pine trees. The basidiospores germinate on wet needles and enter through the stomata.

The only symptom initially is a yellow to reddish spot at the site of infection. Infected needles turn yellow and drop prematurely, but often not before the fungus has grown down the needle and entered the twig. This occurs late in the fall or early in the spring of the next year. About 12 to 18 months after infection, the fungus has grown from the needle into the bark of the stem, where it produces spermagonia. During the first or second year of infection, the spermagonia produce aecia. The aecia produces powdery bright yellow aecidiospores that may live for many months. The aecidiospores are carried by the wind to Ribes bushes, completing the cycle.

Control

1. Remove susceptible *Ribes* species and infected plants.
2. Plant only disease-free resistant varieties of *Ribes* approved by the Department of Agriculture. Some examples of resistant varie-ties of Black currant are Consort, Crusader, Coronet, Ben Sarek, and Ben Nevis. Red currants and gooseberries are not affected by law and are legal to plant.
3. There are no fungicides labeled on currants and gooseberries for control of white pine blister rust.

Chapter 10

Disease Management

Each year plant diseases caused by fungi, bacteria, plant-parasitic nematodes, viruses and air pollution result in some losses in corn production in North Carolina. In 1988, the last year statistics were compiled for North Carolina, plant diseases caused by plant pathogens and air pollution suppressed corn yield by about 10 percent. A number of diseases affect corn, with individual fields often suffering severe losses.

All parts of the plant may be attacked — the ears, leaves, stalks, and roots — at various stages of development. In many instances diseases result in lower yields, but also reduce the value and quality of the grain and may increase harvesting costs when affected plants lodge.

PATHOGENS

Plant diseases, with the exception of air pollutants, are caused by microscopic organisms referred to as plant pathogens. There are several categories of these organisms that may be damaging to plants.

- *Fungi* are multicellular filamentous organisms that are usually microscopic, but sometimes form fruiting bodies visible to the naked eye. Fungi typically produce spores that may infest crop debris, reside on seeds, or remain in the soil. Fungal spores can also serve as a means for long distance spread of these organisms.

- *Bacteria* are microscopic single celled organisms and are typically spread by splashing rain or insect vectors.
- *Viruses* are sub-microscopic and generally must be carried to and injected into the plant by plant feeding insects or nematodes.
- *Nematodes* are round worms that reside in the soil. A few species actively attack the roots of plants and may cause yield losses. Often times plant-parasitic nematodes provide a route of entry for pathogenic fungi and bacteria to enter the plant.

Plant diseases can also be classified as to how they move about, where they reside, and survive. Organisms that infest seed are referred to as being seedborne, organisms that typically reside in soil are classified as soilborne, and organisms that move through the air are referred to as being airborne. These distinctions are important since they often determine the management measures that need to be taken to prevent yield loss.

- *Soilborne diseases* are typically chronic problems requiring an ongoing management programme, that may include rotation and (or) resistant hybrids.
- *Seedborne pests* can often be eliminated through seed treatments, sanitation, or the use of disease free seed.
- *Airborne pathogens* of corn are best managed through the use of resistant hybrids and sometimes cultural practices.

Insect management may aid in suppression of disease caused by bacteria or viruses, but chemical insect control in field corn is rarely economical and resistant hybrids are the most effective means of preventing these diseases. Good cultural practices, including insect management, are especially important in maintaining the integrity of the corn ears, thus preventing losses due to fungal toxins.

Management Principles

For disease to occur three ingredients are essential. You must have a susceptible host, the pathogen must be present, and the environmental conditions for pathogen development on the host must coincide. This concept is referred to as the disease triangle. Because disease is progressive, a fourth factor, time, distinguishes biotic diseases caused by the various organisms versus injury caused by lighting or hail.

The key to managing diseases in corn is the elimination of one of these key factors. The use of resistant hybrids removes the host from the pyramid and thus eliminates disease. Where soilborne plant pathogens are concerned the pathogen may always be present and long-term management strategies must be used such as rotation to reduce the populations of these organisms.

When rotation is impractical, the use of a fungicide or nematicide limits development of the fungus or nematode, and the associated disease is removed or suppressed. Various plant pathogens are favored by different environmental conditions. For example: corn leaf, ear, and stalk diseases generally are favored by warm, wet weather.

Most root rots, such as Pythium root rot, are more severe in wet than in dry soils, but charcoal rot is favored by hot, dry conditions. Seedling diseases are favored by cool soil temperatures that delay emergence and growth of seedlings. Diseases of corn seedlings tend to be more severe when planting time is unusually wet. Symptoms of nematode damage are more pronounced in sandy soils, that can support large nematode populations.

During dry weather the effects of root damage caused by nematodes are apparent as increased plant stress. In some instances we can modify the environment to eliminate the disease. Irrigation may reduce crop stress, which will suppress the amount of disease caused by pathogens that attack stressed plants. Planting on beds places the seedlings in dryer conditions if fields receive heavy rains. Because of the temperature and moisture requirements of different plant pathogens, the arrival of conditions favorable for plant growth may retard or end the progress of the disease in time, and the crop will recover.

There are no completely effective measures for controlling all corn diseases; however, losses can be minimized by following certain recommended practices. Growers should become acquainted with disease problems and factors affecting the severity of the disease.

Disease Management Tactics:

- Rotate crops.
- Destroy crop residue.
- Plant resistant varieties.
- Observe proper planting dates.
- Fertilize properly.
- Harvest at proper time.
- Store corn properly.
- Treat seed.
- Control nematodes with nematicides according to recommendations.
- Use best management practices to minimize other pest problems.

Diseases

Seed Rots and Seedling Blights (caused by species of *Fusarium, Stenocarpella, Pythium,* and other fungi). Germinating corn kernels may be attacked by a number of soilborne or seedborne fungi that cause seed rots and seedling blights. The terms "preemergence" and "postemergence damping-off" are often used to specify the affected growth stage. These diseases are more prevalent in poorly drained, excessively compacted, or cold, wet soils.

Planting old or poor quality seed with mechanical injury to the pericarp will increase seed rot and seedling blight, as will planting seed too deep in wet, heavy soils. Hybrids differ in genetic resistance to the fungi that cause seed rot and seedling blight. Seed treatment with a good fungicide is an important method for control of these fungi.

Southern Corn Leaf Blight (caused by the fungus *Bipolaris maydis [Helminthosporium maydis]).* Southern corn leaf blight occurs worldwide, but is particularly damaging in regions of warm, moist weather. Lesions on the leaves caused by the fungus are elongated between the veins, tan, up to one inch long, with limited parallel margins and buff to brown borders. The fungus overwinters on corn debris in the field. Thus, rotation and destruction of residue will reduce losses due to this disease. Resistant hybrids are also available.

Northern Corn Leaf Blight (caused by the fungus *Exserohilum turcicum [Helminthosporium turcicum]).* Symptoms of this disease are long elliptical, grayish-green or tan lesions ranging from 1 to 6 inches in length, developing first on lower leaves and later causing severe damage to the upper leaves under moderately warm and moist weather conditions.

This disease is favored by somewhat cooler weather than southern leaf blight and has been quite severe in the mountain counties. Northern corn leaf blight can cause premature death and gray appearance of foliage that resembles frost or drought injury. As with southern corn leaf blight, control is by rotation, destruction of crop debris, and use of resistant hybrids. There are at least three pathogenic races of the fungus, but moderate to good resistance is available to all of them.

Anthracnose (caused by the fungus *Colletotrichum graminicola).* Symptoms of this disease vary widely, depending on the hybrid, age of the leaf, and environment. Small, oval to elongate, water-soaked spots first appear on the leaves at any stage of growth. The spots may enlarge up to one-half inch long and become tan at the center with red, reddish-brown, or yellow-orange borders. The lesions may grow together, blighting the entire leaf. Leaf symptoms are most common early in the season on the lower leaves and late in the season on the upper leaves.

Lesions on stalks usually appear initially as black linear streaks under the epidermis. On susceptible plants the lesions may develop into large oval, black areas measuring 1/2 to 1 inch, or larger. In severe infections, large areas of the stalk may be blackened. When the infected stalks are split, a mottled brown discolouration may be seen, particularly at the nodes. This discolouration may be present even when lesions are not apparent on the surface of the stalk. It is common with anthracnose for the upper 1/3 of the plant to prematurely die. Anthracnose is a very important cause of lodging in North Carolina.

Anthracnose is favored by warm, moist conditions during the growing season. Plants are most susceptible in the seedling stage and later as they approach maturity. There is a wide range of susceptibility in hybrids. The fungus over-winters on plant debris left above ground. Thus, control of this disease is based upon crop residue destruction, rotation, and use of tolerant or resistant hybrids.

Southern Rust (caused by the fungus *Puccinia polysora).* Southern rust can be recognized by the bright orange or golden brown, circular to oval pustules, which give a rusty appearance to the leaves. The pustules are about the size of a pinhead and are filled with powdery masses of orange spores, which can be rubbed off. These spores are readily dislodged and blown about in the wind. The spores can survive and infect plants after being transported hundreds of miles by the wind.

The southern rust fungus has no known means of survival in the absence of living susceptible plants. During the winter months it is limited to tropical areas where corn is grown year round. The extent to which it spreads into temperate areas depends upon weather patterns and the susceptibility of the corn along the path of spread.

Southern rust is favored by the warm, humid conditions found in many lowland tropical areas where corn is grown. However, even in those areas, corn with good resistance suffers little or no damage. In temperate areas less ideal for the growth of the fungus, damage can occur in corn hybrids that lack good resistance.

Since southern rust cannot survive the winter in North Carolina, the initial infections must result from spores blown into North Carolina from the south. The fungus can multiply very rapidly on susceptible corn, and the amount of damage that occurs depends upon how early the first spores arrive. Epidemics may result from unusual weather patterns that cause mass air movements from the tropics where the rust is present.

Common Rust (caused by the fungus *Puccinia sorghi)*. Common rust occurs in temperate to sub-tropical areas. It differs from southern rust by the darker, more reddish-brown colour of the pustules. Also, pustules of common rust tend to be longer than those of southern rust and they occur more often in scattered clumps on the leaves.

Pustules of southern rust are usually quite uniformly distributed over the surface of the leaf. Common rust is able to survive the winters in temperate areas because it produces teliospores, which are resistant to weathering. These spores germinate in the spring to produce basidiospores. The basidiospores infect wood sorrel (Oxalis spp.) and the spores produced in infections on wood sorrel complete the life cycle of the fungus by infecting corn.

Common rust has been present for many years in all major corn producing areas of the world. It has not been regarded as a major cause of damage in any of those areas. In 1951 in one of the heaviest outbreaks of common rust known in the United States, estimated average losses ranged from less than 1 percent to 3 percent. Resistance and tolerance to common rust are prevalent and effective in corn hybrids throughout the world.

Common Smut (caused by fungus *Ustilago maydis)*. Common smut occurs wherever corn is grown. Losses to smut are generally light, but may be important in some situations, particularly sweet corn. Young actively growing parts of the plant are susceptible to infection. Large galls may appear on stalks at the nodes, on ears, or rarely on tassels. Leaf infections may result in small inconspicuous galls.

On ears or stalks the galls expand rapidly and are covered with a thin greenish-white or silvery-white tissue. As the galls mature, the covering ruptures exposing masses of black spores within. Individual galls on stalks may be up to 6 inches in diameter. On infected ears, a large number of galls originating from individual infected kernels may combine to form the compound gall mass that replaces most of the ear. Some corn plants may form ears in the tassles.

Smut is usually more severe on plants heavily fertilized with nitrogen. The severity is increased by injury from hail, cultivators, etc. Control involves avoiding highly susceptible varieties, avoiding mechanical injury to plants during cultivation and spraying, and providing well-balanced soil fertility.

Gray Leaf Spot (caused by the fungus *Cercospora zeae-maydis).* The fungus can infect leaf blades and, to a much lesser extent, leaf sheaths. The gray or pale brown lesions are long and narrow with parallel sides delimited by leaf veins. The ends are usually blunt, giving the lesions a long rectangular shape. Lesions commonly are about 1/4 inch wide by about 1 inch long. When the disease is severe, lesions merge into long stripes. Eventually the entire leaf may be killed.

Gray leaf spot has caused moderate to severe damage to corn in the mountain valleys of the Appalachian region. In North Carolina, the disease is most severe in the mountains and western piedmont, but it has become common in the coastal plain and tidewater in recent years.

The gray leaf spot fungus survives the winter as resistant mycelium in corn debris left in the field. The disease is usually more severe in no-till planted corn without rotation. Thus, rotation, debris destruction, and resistant hybrids offer the best methods of controlling this disease.

Brown Spot (caused by the fungus *Physoderma maydis).* Brown spot is favored by high temperatures and high humidity. It attacks leaf blades, sheaths, and stalks, producing small, reddish-brown to purplish-brown spots which may merge together to form large brown blotches. Weakened stalks frequently lodge and leaf sheaths may be reduced to shreds. Good cultural practices and the use of tolerant varieties offer the best control.

Stalk Rots (caused principally by the fungi *Stenocarpella zeae* and species of *Fusarium* as well as *Colletotrichum graminicola).* Stalk rots are present each year and may cause considerable damage, particularly if abundant rainfall occurs during the latter part of the growing season. Stalks previously injured by cold, leaf diseases, or insects are especially susceptible to attack by these fungi.

Diseased stalks ripen prematurely and are subject to excessive stalk breaking. Stalk rots not only add to the cost of harvesting but also bring the ears in contact with the ground, increasing their chance of rotting.

Charcoal rot (caused by the fungus *Macrophomina phaseolina*). Charcoal rot is a destructive disease of corn, soybean, cotton and many other crops. Charcoal rot becomes most evident with the onset of hot dry weather. It may cause a stalk rot, stunting, and death of the corn plant. Symptoms are silver to black discolouration of the stem tissue when the stalk is cut open.

This disease is often considered to be a stress related. Typically when this disease occurs in North Carolina soil fertility and pH are at very low levels. Although the fungus typically survives in the soil, rotation is not generally an option since most crops are susceptible to this disease. Reductions of nutrient and water stress are the principle means of control. Hybrid resistance has not been documented.

Ear and Kernel Rots (caused by species of *Stenocarpella, Fusarium, Aspergillus,* and many other fungi). Some of the same fungi that cause stalk rots of corn also also cause ear and kernel rots. Ear and kernel rots are most serious with warm, wet conditions at harvest time. Severe infection not only reduces yield but also lowers the quality and grade of the grain produced.

The two principle ear and kernel rot fungi found in North Carolina are Aspergillus and Fusarium.

Red ear rot is caused by the fungus Fusarium graminearum (*Gibberella zeae*), and also causes a stalk rot of corn and head scab in wheat. The fungus may cause a reddish discolouration of the cob and kernels. Red ear rot caused by *F. graminearum* is favored by warm wet weather after silking. Disease tends to be worse when corn is grown without rotation or after wheat, as this pathogen also infects wheat. It may be worse when corn is grown in reduced tillage situations.

Fusarium moniliforme is another species of *Fusarium* that causes a kernel rot, but the mycelium of the fungus is typically white to salmon colour. Kernels infected with Aspergillus usually have a greenish to gray colour.

Mycotoxins in Corn

Toxic metabolic by-products of fungi, known as mycotoxins, have received considerable attention during the past several years. Aflatoxin, produced by the fungus *Aspergillus flavus,* has been considered to be the most serious problem in North Carolina in recent years. The detection of aflatoxin in corn can result in a reduced price for the grain or even rejection.

The concentration of aflatoxin in corn for interstate trade is regulated at 20 parts per billion (ppb) by the Food and Drug Administration (FDA). Another class of mycotoxins are referred to as fumonisins. These toxins are produced by the fungus *Fusarium moniliforme* and are quite common in corn produced in North Carolina. It appears likely that fumonisin levels will be regulated in the near future. Corn shipped to Europe will probably be monitored for levels of fumonisin as well as aflatoxin. Allowable levels of fumonisin in corn have not yet been established at this time.

Mycotoxins are known to cause serious health problems in animals including reduced weight gain, capillary fragility, reduced fertility, suppressed disease resistance, and even death. No animal is known to be resistant, but in general, older animals are more tolerant than younger animals. Mycotoxins have been implicated in deaths from acute toxicoses in young animals, particularly poultry, as well as several animal health problems, including reduced fertility and growth rate.

Both *Aspergillus flavus* and *Fusarium moniliforme* are widely distributed in nature and are favored by high temperature. Temperatures ranging from 80 to 100 degrees F and a relative humidity of 85 percent (18 percent moisture in the grain) are optimum for fungal growth and toxin production. Growth of these fungi does not occur below 12 to 13 percent moisture in the grain.

Aflatoxin contamination is higher in corn that has been produced under stress conditions. Thus, drought, heat, insect, and fertilizer stress are all conducive to high levels of aflatoxins. Factors that influence fumonisin production in corn are not well understood at this time. Certainly, insects provide an avenue of infection for both *Aspergillus and Fusarium*. High rainfall and humidity at silking may increase infection of corn kernels by *Fusarium spp*. Hybrids genetically engineered to resist insects have been shown to have lower levels of fumonisin. Therefore, in order to minimize the level of mycotoxins, the following practices should be followed:

- Use recommended production practices.
- Plant early.
- Irrigate to reduce drought stress.
- Harvest early.
- Avoid kernel damage during harvest.
- Dry and store corn properly.
- Keep storage facilities clean.

NEMATODES

Nematodes attack corn roots, thereby limiting their development and restricting the uptake of water and nutrients. Thus, affected plants are stunted and appear deficient of nutrients. Since nematodes do not occur at a uniform population density throughout the field, stunted plants likewise are not uniformly distributed.

They often appear in roughly circular areas in the field. Nematode damage occurs most often when the preplant densities of certain nematodes are high and corn seedlings get off to a slow start because of unfavorable growing conditions. Damage to corn from plant-parasitic nematodes is most severe in the coastal plain area. The two most damaging nematodes on corn in North Carolina are the stubby-root and sting nematodes.

Stubby-root *(Paratrichodorus minor)*. The stubby-root nematode does not enter the roots of corn plants, but remains outside the roots and feeds on the growing root tips. Their feeding prevents the further development of the root tip, resulting in short, stunted or stubby roots. The damage to the root system by stubby-root nematodes resembles that caused by several herbicides.

A plant heavily parasitized with these nematodes is stunted, turns yellow, often exhibits magnesium deficiency, and produces a small ear. Since these nematodes are so widespread in the coastal plain area, they may very well be the most damaging nematodes on corn in North Carolina.

Sting *(Belonolaimus longicaudatus sp.)* - The sting nematode feeds on roots from the outside without penetrating or becoming attached to roots. They feed at root tips and along the sides of succulent roots. Injured roots show blackened, sunken dead areas along the root and at the root tip.

These areas may girdle the root causing it to die. Sometimes the damage done to young plants is quite severe and infected plants may obtain a height of only 8 to 10 inches. Sting nematode is found in soils that contain at least 80 percent sand. This nematode, especially when combined with the stubby-root nematode, causes severe yield losses.

Columbia lances *(Hoplolaimus Columbus)*. The Columbia lance nematode can be damaging to corn, especially if numbers are high and poor conditions for early corn growth occur. The Columbia lance nematode is damaging to corn, whereas a related species of lance nematode, *Hoplolaimus galeatus*, does not generally affect corn.

Currently, the Columbia lance nematode is restricted to sandy soils in parts of the southeastern North Carolina coastal plain, whereas the common lance nematode *(Hoplolaimus galeatus)* is found in many parts of the state. Lance nematodes feed on the root surface but may also penetrate the root system causing internal damage. Columbia lance nematode can be extremely damaging to cotton and soybean, but usually causes only slight-to-moderate damage on corn.

Root-knot and Lesion *(Meloidogyne and Pratylenchus* spp.). Most species of root-knot and lesion nematodes will reproduce on corn. Ordinarily, corn is very tolerant of these nematodes, but may be damaged if populations are very high.

Nematode Management

In order to determine whether or not a field should be treated with a nematicide to control nematodes, a soil sample should be collected in September-November and sent to the N.C. Department of Agriculture for an assay. Soil cores collected from the corn root zone need to be taken in a zig-zag pattern across the field and mixed in a bucket. Each sample should cover 5 to10 acres, and sections with different crop histories should be sampled separately.

Based on estimates from the NCDA&CS, about a third of the corn acreage in eastern North Carolina should be treated to control nematodes. Where the population density is high enough to justify treatment, a grower can expect an increase of about 20 to 25 bushels per acre. Nematodes are controlled by use of a nematicide, rotation, and crop destruction.

Viruses

There are two major viruses of corn in North Carolina, maize dwarf mosaic virus (MDMV) and the maize chlorotic dwarf virus (MCDV). These two virus diseases can cause serious yield reductions, with reported losses ranging from 5 to 90 percent in some fields. Much of the loss due to these two diseases in North Carolina is confined to the piedmont section of the state, although losses in the coastal plain and mountain areas have been reported. This may be due to two factors:

1. There is less johnsongrass in the coastal plain area.
2. The johnsongrass in the piedmont is infected with the two viruses while the johnsongrass in other sections of the state is not as heavily infected, or not infected at all.

The two viruses are transmitted from infected johnsongrass to corn by insects. MDMV is transmitted by aphids (principally the corn leaf aphid, *Aphid maidis* and MCDV is transmitted by leafhoppers *(Graminella nigrifrons).*

Maize Dwarf Mosaic Virus. Symptoms of MDMV first appear on the youngest leaves as an irregular, light and dark green mottle or mosaic which may develop into narrow streaks along veins that appear as dark green "islands" on a lighter green background. As infected plants mature, leaves become yellowish-green.

Plants with these symptoms are sometimes stunted with excessive tillering, multiple ear shoots and poor seed set. Early infection may predispose corn to root and stalk rots and premature death. Symptoms can appear in the field within 30 days after seedling emergence.

Maize Chlorotic Dwarf Virus. MCDV, previously called corn stunt, causes more severe stunting than does MDMV. Infected leaves become yellow, but no mosaic pattern develops. Such leaves usually develop a deep, reddish discolouration later in the season. The internodes of infected plants fail to elongate, resulting in very stunted plants.

Quite often infection occurs late in the season. Thus, the lower portion of the plant develops normally with the upper portion being red and stunted. Infection can result in severe reduction in ear size if susceptible varieties are grown and infection occurs early enough in the development of the plant.

Losses from both MDMV and MCDV can be avoided by growing hybrids that are resistant, or tolerant, to these viruses. There are several hybrids adapted to North Carolina that are resistant to both viruses.

Bacterial Diseases

There are two major bacterial diseases of corn in North Carolina, bacterial leaf blight (sometimes called Stewart's bacterial wilt) and bacterial stalk rot.

Bacterial leaf blight (caused by the bacterium Erwinia *stewartii)* is more of a problem with sweet corn than it is with field corn; however, it can be a problem with certain hybrids. The symptoms are short to long, irregular, pale green to yellow streaks in the leaves. The streaked areas, which die and become straw-coloured, originate from feeding marks of the corn flea beetle.

Sometimes entire leaves die and dry up. When leaves die prematurely, yield is reduced and weakened plants become more susceptible to stalk rots. The bacteria over-winter in corn flea beetles, which also spread the bacteria. Although insect control is important in controlling this disease in sweet corn, it is not a sound practice for field corn producers. Resistance to the disease, which is available in many hybrids, is the preferred method of control.

Bacterial stalk rot (caused by the bacterium *Erwinia* chrysanthemi pv. zeae) can be a problem where overhead irrigation is used and the water is pumped from a lake, pond, or slow-moving stream. Quite often the infection occurs at about ear height, and the upper portion of the plant breaks over due to a collapse of the stalk. Often, an unpleasant odor is associated with this disease. The bacteria usually do not spread from plant to plant, so diseased plants are quite often found scattered throughout the field.

BUNT OR STINKING SMUT

Hosts

The disease is specific to wheat.

Symptoms

Symptoms are very difficult to see in the standing crop. Affected plants are often slightly stunted and have 'squat', dark grey-green ears with slightly gaping glumes. The flag leaves of affected plants sometimes show yellow streaks. If infected ears are broken open they are found to contain, in place of true seeds, seed-like 'bunt balls' each containing millions of greasy, black, foul-smelling spores.

In severe cases, the whole field may smell of rotting fish. In wet weather conditions the ears may appear to be covered in a black ink-like substance as the spores are released and run out of the protective glumes onto the ear and stem.

Many of the bunt balls within the ear are broken open during combining, the spores are released and contaminate grain in both the combine and during subsequent movement and storage handling. In severe cases, the grain can be so heavily contaminated that it is rendered unusable. The spores remain on the seed surface, being usually concentrated in the brush hairs at the end of the grain. In severe cases the grain assumes a dark appearance - often with an obvious foul smell. Older grains which have been stored for some time may lose the obvious smell but are still dark in appearance.

DWARF BUNT

Hosts

This disease is not yet known in the UK but has become of interest in recent years because of reports of long term survival of common bunt in soil. Dwarf bunt can survive for long periods in soil so that early reports of long-term survival of *T. tritici* raised debate about the possibility of hybrids between *T. controversa* and *T. tritici* existing. Dwarf bunt occurs in Canada, North and South America and many parts of Europe and Asia including Austria, Bavaria, Bulgaria, Czech republic, France, Hungary, Iran, Iraq, Italy, Poland, Romania, Russia, Slovakia, Turkey, Sweden and Switzerland.

In general it can be found in most areas where winter wheat is subject to prolonged snow cover through the winter. Unlike common bunt, dwarf bunt is capable of attacking rye and winter barley as well as a range of grass species.

Microscopic examination of teliospore morphology is the primary means of distinguishing species of *Tilletia*. However, because there is a wide range of teliospore characteristics in *T. controversa* and other *Tilletia* species. Spore morphology alone is not a suitable method to identify *Tilletia* species. This identification problem fuels the debate about the possible existence of hybrids between *T.controversa* and *T. tritici*.

Symptoms

The symptoms of dwarf bunt are very similar to common bunt except that affected plants are stunted, to between half and quarter their normal height. Affected ears tend to have a more ragged appearance than with common bunt. Yellowing and flecking of leaves of affected plants is also occasionally

ERGOT

Hosts

Ergot is a very common disease of grasses and cereal crops. It is not truly a seed-borne disease as it is not carried on or in seed. It is however, spread by ergots as seed contaminants and is therefore included here.

Symptoms

The causal fungus only attacks the ear, replacing the grain in a few spikelets by a hard, purple-black sclerotium, known as an ergot. Such ergots can be very large, up to 2 cm in length, and are very obvious in the standing crop and in contaminated grain samples.

FLAG SMUT

Hosts

The disease is known to infect wheat and many grass species but the strain(s) that affect wheat are specific to wheat.

Symptoms

Affected plants are severely stunted. Excessive tillering is common and often the ears fail to emerge, remaining within the boot. Plants show long dark grey to black streaks on the leaf blades and leaf sheaths. The streaks eventually erupt, giving the leaves a ragged appearance and exposing the black teliospores which are then dispersed, giving the plants the appearance of being covered in soot.

FOOT ROT

Hosts

Foot rot due to C. *sativus* is traditionally a disease of hotter climates than that of the UK.

Symptoms

The fungus causes disease symptoms similar to *Fusarium spp*. Seed-borne infection can result in seedling death although infected plants usually grow to maturity. Affected plants show brown spotting on the lower leaves and, if severely affected, can show stem-base rotting and poorly filled ears. This severe symptom is very rare in the UK.

FUSARIUM DISEASES OF WHEAT

Hosts

Wheat, barley, oats, rye, grasses and numerous other species.

Symptoms

These fungi form a complex of diseases on seeds, seedlings and adult plants. Seed-borne *Microdochium nivale* can cause a seedling blight resulting in seedling death and thinning of plant stand. This is particularly likely in late sown crops where seedlings emerge slowly.

Brown lesions on cereal stem bases, which are frequently found in the spring, may be a result of *Fusarium* infection. *Fusarium* is less aggressive than eyespot and consequently lesions are often restricted to the outer leaf sheath. *Fusarium* infection can often appear as a secondary invasion following damage from some other source.

Fusarium lesions are are often found starting at a split in the leaf sheaths at the stem base where crown roots split the leaf sheath when emerging. This infection can then spread up the leaf sheath causing long dark brown streaks at the stem base. Plants affected by late frosts typically become colonised by *Fusarium* - which is often wrongly blamed for the death of the tiller.

Many of the *Fusaria* cause a range of symptoms - often termed 'ear blights'. *Fusarium culmorum* infection frequently results in the whole or part of the ear becoming bleached. This symptom is most commonly seen when ears become infected during the early flowering stages. Later infections may result in infection of the grain but without obvious bleaching of the ears.

The most commonly seen symptom in the UK is the dark brown staining of the lower nodes. On older plants *Fusarium* infection can produce a true foot rot, where the stem base becomes brown and rotten, resulting in lodging and whiteheads. This symptom is rare in the UK, although it can be found in very dry seasons. It is very common in the USA in areas where crops are exposed to very severe drought conditions.

LOOSE SMUT

Hosts

The disease is specific to wheat.

Symptoms

Loose smut is easily recognised at ear emergence as the each grain is usually completely replaced by a mass of black fungal spores. Partly affected ears are sometimes seen. The spores are released as soon as the ear emerges, leaving only the bare remains of the ear rachis. Because the blackened ears are so very obvious in the crop at ear emergence the disease appears to be very severe, even at very low incidence levels.

Septoria Seedling Blight

Hosts

The disease is specific to wheat.

Symptoms

Although more usually associated with necrotic blotching of leaves and glumes, *S. nodorum* can cause pre or post-emergence seedling blight in cool wet soils.

Fusarium seedling blight

Hosts

The most common symptom of a serious attack of Fusarium seedling blight is very poor plant establishment, i.e a very thin crop. This occurs because of pre-emergence and post-emergence seedling blights - the most important phase of the disease.

Root rotting, brown foot rot and leaf blotching (usually following some other form of damage), glume blotch and ear blights can all be caused by true Fusarium species. Severe foot rot, usually affecting plants under stress (especially moisture stress) can result in premature ripening and 'whiteheads'. Under prolonged snow cover *M.nivale* can cause 'snow mould'.

SEED BORNE PATHOGENS

Pathogens Associated to the Seed

As an introduction we will remind some aspects of the morphology and anatomy of the seed, in order to understand in a better way the effect pathogens have in the seed and later, the plant originated from it.

The Seed

The seed consists of three basic parts:

a. Embryo,

b. Storage tissues

c. Seed coat.

a. The embryo has a reproductive function, being capable of initiating cellular divisions and growth. It is the most important part of the seed. It is an axis that originates growth in two directions, with the objective of originating a root and a shoot. Usually, the embryo is very small, compared to other parts of the seed. It is like a miniature plant. It is consists of a radicle, plumule, one or two cotyledons, and hypocotyl or epicotyl, depending on the type of plant.

b. Energy storage is localized in the cotyledons, in the endosperm or the perisperm. Cotyledons originate from the zygote and are part of the embryo. In many species, storage substances are localized in the cotyledons, and the embryo develops absorbing all the endosperm.

c. Episperm is the seed cover, consists of two layers, the testa and the endopleura. The outer layer is the testa; it can be stony, leathery, membranous or fleshy. Over the testa we can recognize: the hilum, scar or point of attachment of the seed to the funiculus, water penetrates easily through it; Micropyle, the point upon the seed at which was the orifice of the ovule through which the pollen tube enters; raphe, suture originated by the part of the funiculus that is fused along the side of the ovule.

The endopleura is the inner layer; it is thin and generally whitish. Teguments, testa or protective covers delimit the seed. They are formed by one or more layers of cells originated from the ovule integuments and sometimes from the pericarp made from the walls of the ovary. Regarding the healthiness or pathology of the seed we will consider three aspects that allow us to understand processes associated to seed-borne diseases: infection of the seed, infection of the plant and steps that can be taken to reduce the damage caused by this relationship.

Seed Infection Mechanisms

The area of science that studies the relationship between pathogens and seeds is Seed Pathology. It does not only identify the pathogens, it also includes the role of the seed as source of inoculum, the survival of the pathogen and the actions taken to control the pathogens associated to it. It uses the knowledge of General Pathology, Microbiology and Seed Analysis.

To obtain a perspective of seed borne diseases, seed borne microorganisms can be considered fewer than four classes. The first consists of pathogens for which the seed is the main source of inoculum; when seed infection is controlled, the disease is controlled. An example would be lettuce mosaic virus. For these pathogens, the importance of seed-borne inoculum has long been recognized, and control practices have been developed.

The second class consists of important pathogens in which the seed borne phase of the disease is of minor significance as a source of inoculum. Examples are those in which the crop residues in the field were the major source of inoculum. The third and largest group of seed borne microorganisms consists of those that have never been shown to cause disease as a result of their presence on seeds.

The fourth class is a group of microorganisms that can infect the seed either in the field or in storage and reduce yield and seed quality. Examples of field fungi are *Diplodia*, *Fusarium*, *Cladosporium*, etc. The storage fungi *Aspergillus* and *Penicillium* can invade most types of seeds under high-moisture storage conditions.

The process of seed infection is influenced by the conditions under which the crop grows. Between the facts that influence in the process of infection are: the host and its genotype, the pathogen and its genotype end environmental facts.

An infected seed will not always an infected seed will be the cause of the infection of the infection in the plant which originates, so, it is to say that the infection caused from an infected seed is an exception, not something to happen generally.

There are two situations, Systemic infection of the seed and Contamination or Infestation of the seed.

Systemic Infection of the Seed

The establishment of a pathogen in any part of the seed is refereed as seed infection. It can be systemic, by the vascular system or plasmodesmata or directly by natural or artificial wounds. The same pathogen can infect the seed using one or more of these mechanisms.

For example: *Xanthomonas campestris* pv. *phaseoli* can infect seed through the vascular system, by natural openings (From the pod suture goes to the funiculus, then to the raphe and tegument, or it can also happen thought the micropyle).

Systemic Infection through Flowers, Fruits or Funiculus

Most of the systemic seed-borne bacteria and fungus reach and infect the embryo through the flower or from the peduncule of the fruit, via funiculus. Viruses go to the embryo from the systemically infected mother plant and the infected or contaminated pollen. They rarely reach the embryo during the formation of the seed or the embryo itself.

Examples of some infections that occur through the vascular system are: Some *forma specialis* of *Fusarium oxysporum* in pumpkin, pea and tomato; *Plasmopara halstedii* in sunflower; *Septoria glycines* in soy; *Verticilium dahliae* in spinach and sugar beet; Certain pathovares of *Xanthomonas campestris* in bean, cabbage, rice and sweet pepper; and *Pseudomonas syringae* pv. *Lachymans* in cucumber.

Penetration through the Stigma

During the infection, some pathogens follow the same path as the pollen grains do. Spores of some fungus reach the stigma and germinate, producing an hypha that reaches the ovary through the style, where they can stay as a dormant mycelium until seed germination.

For example: *Ustilago nuda* and *U. triticci* in barley and wheat, and *Alternaria alternata* in sweet pepper. Viruses can infect through infected pollen, the male gamete carries the virus, when joining the ovule it generates an infected embryo. If both, the male and female gamete are infected they can even produce an infected endosperm.

Penetration through the Wall of the Ovary or Immature Seed Covers

Some fungi, like *Ustilago nuda* and *U. triticci* penetrate through the wall of the ovary as a result of the germination of the Teliospores on the stigma or the wall of the ovary. The pro-mycelium goes through the wall and other tissues until it reaches the embryo. In some other cases, penetration occurs through breakages on the testa, establishing itself in the endopleura or the endosperm.

In fleshy fruits, like cucumber, melon, eggplant, tomato, sweet pepper and others, contamination can occur directly through the funiculus or in the tegument, during the process of seed formation. Examples of this are *Colletotrichum lagenarium* in watermelon and *Rhizoctonia solani,* when it invades fleshy fruits, it is capable of infecting from the placenta and penetrate to the developing ovule or seeds that are still in its formation process and have not lignified its cover.

Penetration through Wounds and Natural Openings

Natural openings like the hilum and the micropyle or wounds generated during the thresh are spots where pathogens like *Xanthomonas campestris* pv. *phaseolicola* in bean and *Pseudomonas syringae* pv. *lachrymans* in cucumber.

SEED CONTAMINATION OR INFESTATION

Contamination or infestation refers to the passive relationship of a pathogens and seeds. The pathogen itself or parts of it can stick or can get mixed with the seeds during any of the processes during seed recollection: harvesting, extraction, thresh, selection and packing.

Pathogens that Stick to the Surface of the Seed

Pathogens that stick to seeds during harvest or postharvest do so by their spores (Clamidospores, Oospores,

Teliospores, Uredospores), bacterial cells and in some cases, virions. Spores of the following fungi can be carried on seed coat surfaces: *Alternaria brassicae and A. brassicicola* in crucifers; *A. longipes* in tobacco; *A. radicina* in carrot; *Ascochyta pinodella* in pea; *Drechslera sorokiniana* and *D. oryzae* in rice; *D. avenae* in oats; *Fusarium oxysporum* f. sp. *callistephi* in China aster; Sclerotia of *Rhizoctonia solani* in eggplant, pepper, and tomato; *Tilletia caries, T. foetida, T. contraversa,* and *Urocystis agropyri* in wheat.

A number of bacteria, such as the following, contaminate seed surfaces:

Corynebacterium flaccumfacjens pv. *flaccumfacjens* in bean, *Pseudomonas syringac* pv. *phaseolicola* in bean, *P. syringae* pv. *lachrymans* in cucumber, *P. syringae* pv. *tomato* in tomato, *C. rnichiganense* pv. *michiganense* in pepper, *Xanthomonas campestris* pv. *campestris* in cabbage Also some viruses as Tobacco mosaic virus, Tomato mosaic virus, Pepper Mosaic virus.

ACCOMPANYING CONTAMINATION

This type of contamination refers to physical mixing of the seed with pathogen's propagation organs like the sclerotium, nematode's galls, contaminated plant parts or soil particles containing pathogens.

Structures of the Pathogens

Some fungi produce resistance structures named sclerotiums that are made of compacted mycelium. Under certain conditions of temperature and humidity it can germinate, it is common that this occurs together with the seed, producing its infection, by direct penetration or by the spores generated by the fructification organs (Apothecium or Perithecium).

These spores are carried by the wind and taken to susceptible tissues of the plant, generally, some parts of the flower. Sclerotiums have many shapes and during thresh can be easily included with the rest of the seed. Some examples of this are *Sclerotinia sclerotiorum* in horticultural, grain and flower crops, *Sclerotinia cepivorum* in garlic and *Claviceps purpurea* in barley.

Mix with infected Plant Parts

Infected plant's parts and residues can carry fructifications or spores of fungi and bacteria. This situation is particularly common in cereal, forage and oily crop's seeds. Examples of this are *Septoria nodorum* in wheat, *Puccinia malvacearum* in hollyhock (malvarrosa), *Colletotrichum trifolii* in clover, *Erwinia carotovora* pv. *carotovora* in tobacco, *Pseudomonas syringae* pv. *phaseolicola* in bean and *Sclerotinia sclerotiorum* in sunflower, soy and peanut.

Soil

Seeds can be mixed with contaminated soil and, carry micro-sclerotiums of *Macrophomina phaseolina* in dirty seeds of kidney bean, *Verticillium oxysporum* f.sp. *phaseoli* in cotton seeds, *Plasmodiophora brassicae* in turnip and *Fusarium oxysporum* f. sp. *phaseoli* in bean.

SEED TRANSMITION

We saw how the seed gets infected or infested. Now we will see how these seeds can produce an ill plant. This is to say, how the inoculum goes from the seed to the plant. We call seed-borne pathogens only those that can produce an infection. Because they have to be distinguished from those that can associate to the seed but do not produce an infection, these ones are called pathogens not transmitted by seed.

The virus that causes curly top in sugar beet can be present in the perisperm but does not cause an infection in the plant. In general terms, infection can be classified into systemic and non systemic. It is systemic when the pathogen introduces itself to the plant when the seed germinates, and develops with it.

Non systemic infection occurs when there is a localized infection caused by the pathogen in the seedling at the stage of pre or post emergency, in this situation there are no systemic symptoms.

Systemic Transmition

This type of infection can be produced by pathogens that are carried with the seed in various parts, like the embryo, endosperm or episperm, or by contamination of the outer portion of this one.

Infection of the Embryo

When the seed germinates, if the embryo is infected, the pathogen initiates its growth together with the plant. Symptoms can show up during different stages of development. In the case of *Ustilago triticci,* the mycelium grows together with the plant and expresses symptoms only at the flowering stage, which are that all the tissues of the spike, exept the rachis, are replaced by spores.

Some pathovares of *Xanthomonas campestris* that infect cabbage, bean or sweet pepper move between the cells of the host until they reach the vascular system and produce symptoms in leaves or stems. Most of seed-borne viruses persist inside the embryo. Its multiplication and movement accompanies the plant during its development, and there can be symptoms at any stage, from the formation of first leaves until flowering or fructification.

Non Embryo Infection

Infection of the episperm (testa and endopleura) occasionally conducts to a systemic infection. Some bacteria, like *Corynebacterium michiganense* pv. *michiganense* in tomato and *Xanthomonas campestris* pv. *campestris* in cabbage, penetrate through stomata of cotyledons, and from there, reach the vascular system, initiating the systemic infection.

Episperm Contamination

In some few cases this type of contamination conducts to a systemic infection. Most of these exceptions are fungi that are highly specialized in their pathogenesis and produce in cereals the so called smuts, rusts or mildews.

In these cases, generally, spores are carried outside the seed, they germinate, penetrate the coleoptile, and start a systemic infection. This can occur directly o by a more complex system of haploid hypha fusion in genera like *Tilletia, Ustilago, Uromyces, Sclerospora, Pernospora and Puccinia.*

Non Systemic Transmition

Non systemic infection is very common, and in the same way as systemic infection, it can come from an infection, outside contamination or by pathogens mixed with the seeds.

Infection of the Embryo

This case is restricted to some pathogenic fungi that maintain itself in the embryo or the episperm as hypha inside the seed. Primary infection starts as injuries in the cotyledons or primary leaves, stems or petiole. Fungi fructifications (Pycnidium, Acervulus) can develop on these organs, under certain favorable conditions (temperature and humidity).

These fructifications produce spores that, with the action of water and wind, disperse the disease to other parts of the plant and other plants. This occurs in *Ascochyta pisi* in pea, *Colletotrichum lindemuthianum* in bean and *C. truncatum* in soy.

Infection of the Episperm

Generally, seeds in which the episperm is infected, do not geminate, or germinate and contaminate the soil. Rarely produce a systemic infection, but can infect the seedling from the outside.

Over the injuries, new inoculum is produced; this one infects other parts of the plant or other plants. *Septoria nodorum* in wheat under high humidity conditions forms Picnidium in the coleoptile, whose spores disseminate the disease to other plants.

Contamination of the Episperm

Contamination outside the seed's testa can produce healthy seedlings, but the inoculum infects the soil and from there it can cause infections at more developed stages of the plant.

Wheat and rice grains that are contaminated with Teliospores of *Neovossia indica* in wheat and *N. torrida* in rice, produce ESPORIDISPORAS in the soil, that can be carried by the wind and infect flowers, originating the smuts.

Accompanying Contamination

The mixture of seeds with sclerotiums of the fungi or contaminated soil particles produces non systemic infections, at any stage of the plant growth and development. It is common for *Sclerotinia sclerotiorum* to produce mycelium from the sclerotiums.

These mycelium can directly infect the plant or produce fructifications that produce spores that can be carried by the wind and infect flowers of different species, like sunflower, soy, peanut, etc.

Prevention of Seedborne Diseases

Prevention of seed-borne diseases is based in two main facts. On one hand there is the usage of material free from pathogens and contaminants that could originate a disease and on the other there are the treatments to eliminate the inoculum in the seed or propagule.

When using material free from pathogens it is important to have adequate pathogen detection methods, this is to say, sensible and specific. No production of health quality seed can be done without valid methods to detect the possibility of contamination. Some treatments can be necessary to eliminate the pathogens from the seeds, or at leafs, maintain an economically viable balance.

Health Testing of the Seed

One of the most relevant aspects about controlled health quality seed is refereed to the detection methods used on every case. ISTA (International Seed Testing Association) appeared in 1924, by was only interested in seed's gene purity and germination aspects. Later then, thanks to the efforts made by Dr. L. C. Doyer publishing the first Manual for Determination of Seed Borne Disease and being the first director of the PDC (Plant Disease Committee), the importance of seed-borne diseases was turned unto first priority.

But detection methods were applied according to the criterion of each laboratory, so their results were not comparative. So, in 1957, PDC established a comparative health testing programme and standardize applied methods. The complexity for comparison of these methods made a considerable step back on the publication of working sheets (protocols). ISTA approved methods became Rules. By 1999 only 64 protocols were being compared, and only 14 of them were accepted as Rules.

As the solutions proposed by ISTA did not satisfy the needs of international seed industry, this industry, in 1994, organized The International Seed Health Initiative for Vegetables (ISHI-Veg) chartered by the vegetable seed industries in The Netherlands and France. The Initiative was soon joined by the seed companies in the United States, Israel and Japan. This group represents the production of over 75% of the worlds vegetable seed supply. ISHI published many protocols, organized by crop. For horticultural crops, the ISHI-VEG, the Seed Health Testing Methods

ISHI-Veg Status

1. An ISHI Validated Reference Method is a method that has been through the ISTA/ISHI comparative testing process and is being published as ISTA working sheet.
2. An ISHI Reference Method is a method that has been through ISHI comparative testing and is under review by the ISTA/ISHI reviewers for publication as ISTA Working Sheet.
3. An ISHI Accepted Method is a method that is commonly used by the seed industry for determining seed health. The method has been published in a

journal, as a working sheet and/or is publicly available for comparative testing and commonly in use by the seed industry. The method is in comparative testing or aspects of the test are being tested for inclusion or deletion from the method.

4. An ISHI Reviewed Method is a method that is publicly available or published, documented and prioritized by the ITG, but not subjected to a comparative test at this time. The method usually is accepted by other agencies or groups.

In 1995, during the second ISTA-PDC symposium it was decided that a Joint ISTA/ISHI Guidelines for Comparative Testing of Methods for Detection of Seed Borne Pathogens would be edited.

Through this combined effort many laboratories, organized under this alignments, the ISTA developed the Manual of Seed Health Testing Methods that has two sections:

I. - Validated Test Methods

II. - Peer Reviewed Methods.

The transition process from working protocol to ruled method is long and laborious, so working protocols can be considered as adequate methods for seed analysis. Nowadays; it is common that technological advances may become obsolete even before they are published.

The protocol to analyze *Xanthomonas campesrtis* pv.*campestris*, which causes black rot in crucifers, was published in 1982, there, it was recommended one minute to extract the bacteria from the seeds to starch culture medium. 2 years later, this time was raised to 2 hours, and other specific culture mediums. In 1987 the protocol was replaced, again in 1996 and at present time again it is under revision.

New Techniques for Seed Health Analysis

Pathogen detection methods have suffered great changes since 1957. Back then it was mainly targeted to fungi, and based exclusively on incubation and identification, or the cultivation of these seeds and counting the percentage of unhealthy seedlings. These analyses were very laborious and required a high infection level in order to be detected.

Nowadays, technological advances have allowed detecting even low infection levels, evaluating 1,000 to 10,000 seeds at once. Specificity of culture mediums, usage of antibiotics and other products has allowed isolating the pathogen from the seed. Together with this, the use of reactives as mono or polyclonal antiserums and molecular markers is also available. The Lettuce Mosaic Virus (LMV), back in 1950 the symptoms were observed over many thousands of seedlings from the seed that were going to be analyzed.

Later, technical advances determined that 30,000 seedlings were needed, and even later, in 1983, an alternative to the *Chenopodium quinoa* test was discovered, the ELISA test.

Watermelon Fruit Blotch caused by bacteria is a severe problem in glasshouse winter production. It was determined that a single contaminated seed in 9,000 of them was enough too widespread the bacteria to the rest of the plants. That is the reason why screening was done over 10,000 seedlings. Nowadays PCR technique is used.

Minimum requirements to Produce basic and Certified Seed

The Organization for Economic Cooperation and Development (OECD) establishes steps to be taken in order to produce basic and certified seed. Between these steps it is emphasized that generations from the mother material to the basic material are strictly limited in number. The number of certified seed generations, after the basic seed, varies from crop to crop and technical conditions.

In both, sexually and vegetative reproduced species, the schemes include similar steps. It generally starts with selection of a plant with genetic and health advantages. If health condition was not satisfying, certain actions should be taken in order to release it from considered pathogens. This Mother Plant is maintained under strict special security conditions in order to maintain healthiness (isolated glasshouses or chambers).

In the Certification Schemes two particularities abut seed borne pathogens should be considered, those transmitted exclusively with the seed and those that can also be transmitted with contaminated soil and plant rests. Diseases transmitted exclusively with the seed can be controlled by using clean healthy seed. This seed can be obtained by Seed Certification Programmes or by treatments that are really effective, like *in vitro* culture, thermotherapy, chemotherapy or a combined action of these, for systemic pathogens.

Tolerance levels in certified seed depend on the multiplication speed and dispersion of the inoculum of the pathogen of the considered disease. If a small amount of inoculum may be epidemic, like Anthracnose and the bean bacterial blight, if levels above zero chemical control may not be enough to stop the disease.

But pathogens whose inoculum multiplies slowly generation after generation, the disease may be evident only several generations later, so, if contamination levels are above zero may be accepted in adequate control methods are applied to the seed. This is what happens in wheat, with the Loose Smut, continuous treatments diminish inoculum levels, and make it perfectly tolerable.

Index

W

Z